Advances in Synthesis, Processing, and Applications of Nanostructures

Advances in Synthesis, Processing, and Applications of Nanostructures

Ceramic Transactions, Volume 238

Edited by
Kathy Lu
Navin Jose Manjooran
Ri-ichi Murakami
Gary Pickrell

A John Wiley & Sons, Inc., Publication

Published by John Wiley & Sons, Inc., Hoboken, New Jersey.
Published simultaneously in Canada.

For general information on our other products and services or for technical support, please contact our
Customer Care Department within the United States at (800) 762-2974, outside the United States at
(317) 572-3993 or fax (317) 572-4002.

Wiley also publishes its books in a variety of electronic formats. Some content that appears in print may
not be available in electronic formats. For more information about Wiley products, visit our web site at
www.wiley.com.

Library of Congress Cataloging-in-Publication Data is available.

ISBN: 978-1-118-27327-2
ISSN: 1042-1122

Printed in the United States of America.

10 9 8 7 6 5 4 3 2 1

Contents

NANOTECHNOLOGY FOR ENERGY, HEALTHCARE, AND INDUSTRY

Preface

There have been extraordinary developments in nanomaterials in the past two decades. Nanomaterial processing is one of the key components for this success. This volume is a collection of the papers presented at three nanotechnology related symposia held during the Materials Science and Technology 2011 conference (MS&T'11), October 16–20, 2011 in Columbus, Ohio. These symposia included Controlled Processing of Nanoparticle-based Materials and Nanostructured Films; Nanotechnology for Energy, Healthcare, and Industry; and Synthesis, Properties, and Applications of Noble Metal Nanostructures.

Nanoparticle-based materials and nanostructured films hold great promise to enable a broad range of new applications. This includes high energy conversion efficiency fuel cells, smart materials, high performance sensors, and structural materials under extreme environments. However, many barriers still exist in understanding and controlling the processing of nanoparticle-based materials and nanostructured films. In particular, agglomeration must be controlled in powder synthesis and processing to enable the fabrication of homogeneous green or composite microstructures, and microstructure evolution must be controlled to preserve the size and properties of the nanostructures in the finished materials. Also, novel nanostructure designs are highly needed at all stages of bulk and thin film nanomaterial formation process to enable unique performances, low cost, and green engineering. This volume focuses on three general topics, 1) Processing to preserve and improve nanoscale size, structure, and properties, 2) Novel design and understanding of new nanomaterials, such as new synthesis approaches, templating, and 3D assembly technologies, and 3) Applications of nanoparticle assemblies and composites and thin films.

We would like to thank all symposium participants and session chairs for contributing to these high-quality and well attended symposia. Special thanks also go out to the reviewers who devoted time reviewing the papers included in this volume. The continuous support from The American Ceramic Society is also grateful-ly acknowledged. This volume reflects the quality, the scope, and the quality of the presentations given and the science described during the conference.

KATHY LU
NAVIN JOSE MANJOORAN
RI-ICHI MURAKAMI
GARY PICKRELL

Controlled Synthesis, Processing and Applications of Structural and Functional Nanomaterials

EFFECT OF ANNEALING AND TRANSITION METAL DOPING ON STRUCTURAL, OPTICAL AND MAGNETIC PROPERTIES OF ZnO NANOMATERIAL

Navendu Goswami
Department of Physics and Material Science and Engineering, Jaypee Institute of Information Technology, A-10, Sector-62, Noida -201307, U.P., India.

ABSTRACT
        The unique optical, magnetic and electronic properties of nanoscale ZnO make them an ideal material for diverse potential applications. We synthesized different shapes and sizes of undoped and transition metal doped ZnO nanostructures. The prepared nanomaterial was systematically studied to probe the effect of size, shape, synthesis process and doping on their structural, optical, electronic and magnetic properties. Here we elucidate the crucial role of defect states in the luminescence of pre- and post-annealed undoped ZnO nanoparticles. This finding paves the way to the process of altering visible luminescence in ZnO nanoparticles without varying their size or shape. Furthermore, we demonstrate the vital role of transition metal doping to alter the shape, size and subsequently optical and magnetic properties of ZnO nanostructures.

INTRODUCTION

        Nanomaterials have attracted immense attention worldwide due to their exotic properties as compared to their bulk counterparts [1-4]. Among various domains of nanoresearch, the control of morphology and structure of nanomaterials has always been one of the most fascinating yet challenging goals since properties of nanomaterials are essentially driven by their shape, size and distribution [5]. Group II–VI nanoscale semiconductors, recognized for their unique optical and electronic properties have been especially investigated because of their vast potential applications in optoelectronic devices, solar cells, biological labeling, and so on [2, 6, 10-12]. In this class of direct bandgap intrinsic semiconductor nanomaterials, functional oxides, e.g. ZnO are the fundamentals of several smart devices such as solar cells, field effect transistors and gas sensors [13-18]. Due to semiconducting and piezoelectric dual properties, the uniqueness of ZnO is recently utilized to devise nanogenerators [19, 20]. Zinc oxide is widely applicable for optoelectronic and luminescent devices due to its remarkable luminescence properties [21]. The fluorescence properties of ZnO are significantly influenced as the dimension of material is reduced. A number of theoretical reports are published to elucidate the size-dependent optical properties of semiconductor clusters [22].

        Since properties of ZnO nanocrystals depend closely on their size, morphology, surface area and activity, novel nanostructures of ZnO were fabricated in diverse shapes such as nanocrystals, nanobelts, nanowires, nanosheets, nanodiskettes and tetrapods [13, 19, 23, 24]. Nanostructured ZnO with different particle sizes and morphologies have been fabricated by several techniques, such as thermal decomposition, sol-gel method, gas-phase reaction, and hydrothermal synthesis [25-30]. Various chemical precipitation techniques have been popularly adopted to synthesize ZnO nanoparticles [31]. Among these synthesis methods, it is well conceived that solution chemical routes are more convenient and less expensive and have general advantages such as superior uniformity and high yield of nanoparticles [32].

        Despite immense research work on doped nanostructured materials, there are few reports on transition metal doped ZnO nanostructures, particularly about their structural and optical properties [33-35]. The doping of nanostructure with transition metal is an effective technique to adjust the energy levels and surface states of ZnO, which can further introduce changes in its physical and optical properties [34]. However, it is still a great challenge to synthesize transition metal doped ZnO nanostructures using a simple process with a low cost and high yield [33]. Among various transition

3

metal doped materials, the growth and study of Ni-doped ZnO nanostructures is of enormous significance due to their potential applications in spintronics, opto-electronics and sensing devices [33-35].

In the present study undoped and Ni doped ZnO nanostructures were synthesized by a chemical method. The structural, electronic and optical properties of undoped and Ni doped ZnO nanostructures were probed employing XRD, TEM, SEM, EDAX, UV-Visible and FL spectroscopy. We have made an attempt to decipher the role of various reaction parameters involved in the growth of nanoparticles. Processing and mechanism of synthesis has been elucidated on the basis of these investigations. The effect of annealing on the structural and optical properties is extensively investigated. Furthermore, the green emission and change in the radiative transitions for unannealed and annealed ZnO nanoparticles are elucidated. Finally we explain the influence of Ni doping on the modification of structural and electronic properties of doped ZnO nanostructures. In order to vividly explicate the aforesaid phenomena, the report is divided in two segments, namely 'A' and 'B'. Segment 'A' deals with the study of unannealed and annealed undoped ZnO nanoparticles whereas segment 'B' contains the report pertaining to investigations on Ni doped nanostructures.

METHODS AND MATERIALS

1. Synthesis Procedure

We synthesized ZnO nanoparticles employing a chemical precipitation technique. In this method, 1M aqueous solutions of zinc acetate dihydrate $\{Zn(CH_3(COO))_2.2H_2O\}$ and sodium hydroxide (NaOH) were prepared in doubly de-ionized water. These solutions were slowly reacted in a vessel to produce the precipitate of ZnO particles.

The basic chemical reaction governing the formation of ZnO is as follows:

$Zn(CH_3(COO))_2.2H_2O + NaOH \longrightarrow ZnO + 2CH_3(COO)Na + 3H_2O$

Earlier it has been confirmed that nanoparticles of different sizes can be synthesized by modification of pH of the system [36]. Since the size of nanoparticles prepared depends upon the pH value of solution, the pH value of the product solution is regularly recorded after mixing the 5ml of each reactant. The pH values of 5.9 and 12.2 respectively symbolize the acidic and basic nature of 1M zinc acetate and 1M sodium hydroxide reactant solutions. For this study we have prepared the 100ml of product solution by slowly mixing the 50ml of each of the reactant solutions. The pH value of product solution is 7.32. ZnO particles, thus formed, are in the form of white coloured precipitate. The precipitate of ZnO was thoroughly washed with de-ionized water to discard the remaining solvents. Finally, the powder of ZnO particles was dried in an electric oven at 100ºC for 12 hour.

Extending the same procedure, the Ni doped ZnO nanostructures (say Ni:ZnO NS) were prepared. In order to obtain 1%, 3%, 5%, 7% and 10% doping of Ni in ZnO, the calculated amount of NiO solution was added into the solution of zinc acetate. In this solution the aqueous solution of NaOH was added. This precipitate of Ni:ZnO nanostructures was thoroughly washed and dried at 200°C for 6 hours. The collected dried precipitate was grinded and annealed at 400° C for 2 hours to further remove the impurities of carbonates etc. Finally, Ni:ZnO powder is finely grinded for various characterizations.

2. Mechanism of Synthesis

Understanding the growth mechanism is critical in controlling and designing the ZnO nanoparticles. In our synthesis method the addition of reactant solutions of zinc acetate and sodium hydroxide initiate the process of precipitation. During the process of precipitation, ZnO nanoparticles are originated from product solution. In this precipitation technique, composition of the solvent is modified in such a way that the ZnO nanoparticle is formed that has a significantly lower solubility than the concentration in solution [37].

The formation of ZnO nanoparticles proceeds step by step from crystalline/amorphous seeds (primary particles) to larger particulates. In order to achieve monodisperse nanoparticles, it is crucial here that the seed formation rate or nucleation rate, respectively, be faster than the growth rate of particles [37]. Due to this reason, the slow mixing of reactants in our methodology is expected to increase nucleation rate and decrease growth rate. In a continuous precipitation process, the particle size distribution as well as the structure of the particulates can be fine turned by process engineering, i.e. by choosing both the appropriate 'flow conditions' and 'particle-particle interactions'. In our synthesis these parameters can be controlled by pH value and rate of reaction [37].

3. Characterization

In order to study the structural, electronic and optical properties of colloidal and powder samples of ZnO were characterized. Colloidal ZnO nanoparticles were obtained by simply dispersing the ZnO powder in de-ionized water or methanol, as per requirement. Furthermore, we have analyzed the annealed and unannealed samples of ZnO nanopowder. Identification of the crystalline phase of ZnO powder was performed using a Philips X'pert Materials Research X- ray diffractometer with a Cu anode, generating Kα radiation at 1.544 A^0 and operating at 35 kV and 30 mA. XRD θ-2θ patterns were recorded in the 2θ range 0-70^0 at the scan rate of 2.4degree/minute. All Size, shape and size-distribution of prepared nanoparticles were studied employing Philips CM-12 Transmission Electron Microscope operated at 100KVA. Optical absorption spectra for the ZnO colloidal particles were recorded using a Perkin-Elmer Lambda-32 UV-Visible spectrophotometer. Fluorescence excitation and emission spectra of annealed and unannealed ZnO nanoparticles were recorded employing Perkin-Elmer LS55 Fluorescence spectrometer. In this monochromator based spectrometer, a high energy pulsed Xenon source is used for excitation. The fluorescence spectra are obtained for ZnO nanoparticles suspended in the methanol and placed in the quartz cuvettes of 1cm path length. The fluorescence excitation and emission data is collected at the scanning interval of 0.5nm.

In order to study the structural, electronic and optical properties powdered phase of prepared Ni:ZnO nanostructures was used. Identification of the crystalline phase of Ni doped ZnO nanopowder was performed using a Philips Analytical X-Ray B.V. Diffractometer. The shape, size and distribution of nanostructures were studied employing JEOL JEM-2100F HR-TEM equipped with an EDAX. The SEM images of nanostructures were also captured employing a Philips XL−20 SEM.

RESULTS AND DISCUSSION

As previously mentioned, we present the results and discussion pertaining to unannealed and annealed ZnO nanoparticles in the segment 'A' and study of 1-10% Ni doped ZnO nanostructures is presented in segment 'B'.

A. Study of unannealed and annealed ZnO nanoparticles

In the following sections, structural and optical properties of unannealed and annealed undoped ZnO nanoparticles are expatiated.

A1. XRD Analysis

The X-ray diffractogram of the powder ZnO nanoparticles is depicted in figure 1(a). The prominent peaks labeled by filled circles at 31.6°, 34.2°, 36.1°, 47.3°, 56.3°, 62.7°, 66.2°, 67.5° and 68.8° correspond to the (100), (002), (101), (102), (110), (103), (200), (112) and (201) planes confirming the formation of hexagonal zinc oxide phase (JCPD 36-1451 for wurtzite zinc oxide) [32, 38]. The formation of nanocrystalline ZnO is reflected through the broadening of the XRD characteristic lines for ZnO. The average size of nanocrystallites determined using the Debye-Scherrer equation ($D = k\lambda/\beta Cos\theta$, k = 0.99 for spherical particles) is 20.5 nm [39, 40].

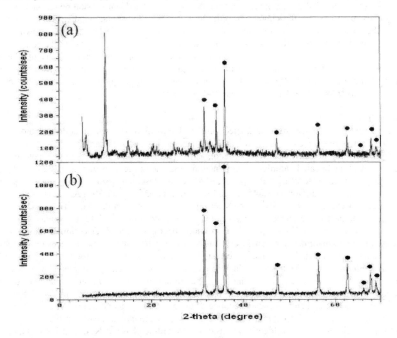

Figure 1. XRD patterns of ZnO powder (a) before annealing and (b) after annealing at 600°C.

The additional unlabeled diffraction peaks in figure 1(a) reveal the presence of carbonate complexes in the sample. These lines clearly indicate that drying of ZnO precipitate at 100ºC is not sufficient to remove the absorbed carbonate in ZnO nanoparticles. XRD result prompted us to anneal the ZnO nanoparticles at 600ºC for 1hr. As the melting point of bulk ZnO is 1975ºC, we have carefully chosen this temperature as it is well below the expected reduced melting point for ZnO nanoparticles [24]. XRD pattern of the annealed ZnO nanoparticles is shown in the figure 1(b).

The absence of carbonate peaks in the XRD data for annealed powder demonstrates the complete removal of carbonates from ZnO nanoparticle sample after annealing. We also analyze that there is no effect of annealing on the crystalline phase of ZnO as the XRD characteristic lines, labeled by filled circles, remain unshifted in pre- and post annealed data. Interestingly the lattice plane (200), which was not distinctly observed in for unannealed ZnO, is prominently distinguished for annealed ZnO. This suggests that annealing helps in improving the crystallinity of ZnO nanoparticles. However, the size of ZnO nanocrystals, as estimated by Debye-Scherrer equation, remains unaffected due to annealing. These results further validate our choice of annealing temperature, 600ºC.

In order to further understand the effect of annealing on the properties of ZnO nanoparticles, we characterize the ZnO nanoparticles before and after annealing treatment.

A2. TEM Analysis

The TEM images of the pre- and post-annealed ZnO nanoparticle are shown in the figure 2. No significant difference is observed in both the images. In both cases nearly spherical ZnO nanoparticles are observed. The smallest size of the ZnO nanoparticles is about 18nm whereas the average size of nanoparticles is around 24nm. The ZnO particles in the size range of 18-42nm can be observed in both cases.

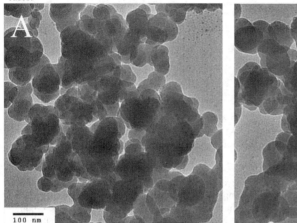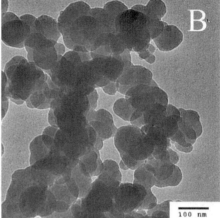

Figure 2. TEM image of ZnO nanoparticles (A) before and (B) after annealing (Magnification: 66kX).

The discrepancy in the size estimation of nanoparticles by different technique can be understood as follows: It is a well established fact that the precision of crystallite size analysis by Debye-Scherrer calculation is, at best, of the order of ±10% [39-42]. Therefore, the actual smallest size of ZnO nanocrystal, as determined by TEM, is 18nm. However the crystal size estimated by the x-ray diffraction, which identifies the underlying lattice planes, is 20.5nm. This value of crystal size is in agreement with the TEM result with the precision limit of ±10%.

A3. UV-Visible Analysis

We have investigated the radiative absorptions in ZnO nanoparticles employing UV-Visible spectroscopy. Figure 3 shows the UV- visible absorption spectrum of the unannealed powder sample. In the figure, one can observe the optical absorption region due to ZnO nanoparticles appearing in the range 310-325nm. This broad region indicates the size distribution of ZnO nanoparticles, as evident by TEM and XRD results too. For the purpose of band gap measurement, the peak position is ascertained by differentiating (i.e. d(Abs)/dE) the experimental plot in the range 250-400 nm to arrive at a peak position of approx 315nm which corresponds to a band gap of 3.93eV. This represents the blueshift in the band gap of ZnO nanoparticles from its bulk band gap energy 3.20eV [32]. The increase in the band gap with decrease in the crystallite size is attributed to size quantization effects.

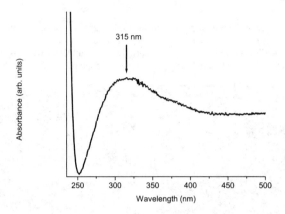

Figure 3. UV-visible absorption spectra of the unannealed ZnO nanoparticles.

A4. Fluorescence Spectroscopy Analysis

Generally, ZnO exhibits a visible deep level emission with a peak in the range from 450 to 730nm [43]. Among the various luminescence properties of ZnO nanoparticles, the green emission around 521nm is usually reported [21]. However, out of various reported emission peaks, the origin of the green emission is the most controversial [21]. Vanheusden et al. elucidated the mechanism behind green photoluminescence and claimed that oxygen vacancies are responsible for the green luminescence in ZnO [44].

In order to probe the possible effect of annealing on the typically reported green photoemission (~521nm) in ZnO nanoparticles, we performed fluorescence spectroscopy. For this purpose, firstly we have recorded the fluorescence excitation spectra to identify the prominent excitation line(s) for ZnO nanopowders. Fluorescence excitation spectra for annealed and unannealed ZnO powders are recorded in the excitation range of 325-500nm for the fixed emission line of 521nm. The fluorescence excitation spectrum of unannealed ZnO nanoparticles is shown in the inset of figure 4. The FL excitation spectrum of annealed ZnO is found to be same as of unannealed one.

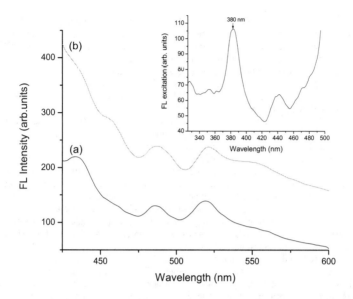

Figure 4. FL emission spectra of (a) annealed and (b) unannealed ZnO nanoparticles excited with 380nm line. An FL excitation spectrum of unannealed ZnO nanoparticles for the fixed emission line at 521nm is also shown in the inset.

It can be seen in figure 4 that a prominent peak at 487nm is present for annealed as well as unannealed ZnO nanoparticles. However, another prominent peak positioned at 519nm for annealed ZnO nanoparticles demonstrates the blue shift in the green emission as compared to that of unannealed ZnO nanoparticles; where the peak is observed at 521nm. In case of unannealed ZnO, two emission humps at 453nm and 551nm are observed. But these emission humps are not seen for annealed ZnO. However, we do observe a prominent peak at 433.5nm for annealed ZnO.

It is evident from the emission spectra that both the unannealed and annealed ZnO nanoparticles synthesized by our method exhibit green emission; however, the process of optical recombination differs in unannealed and annealed ZnO nanoparticles. This indicates that annealing of ZnO nanoparticles not only purges carbonates and moisture from the material but it also influences various channels of radiative recombination.

The optical properties, i.e. absorption and fluorescence spectra of ZnO nanoparticles can be explicated in terms of various radiative transitions occurring within band structure. For wurtzite ZnO, valence band (VB) is predominately O-2p in character whereas Zn 4s – O 2p σ^* interaction is primarily responsible for two lowest conduction band (CB) that span a bandwidth of ~10eV. There is dispersion in the VB due to various factors out of which one reason is mixing of filled Zn 3d and O 2p orbitals [45]. Visible luminescence in ZnO is primarily caused by the transition from deep donor level to VB due to oxygen vacancies and by the transition from CB to deep acceptor level due to impurities and defect states [21]. The crystal structure of ZnO contains large voids which can easily accommodate interstitial atoms and the appearance of blue emission in annealed ZnO nanoparticles at about 433.5nm is ascribed to the formation of Zn interstitial defects [46]. In ZnO, oxygen has tightly bound 2p electrons and Zn has tightly bound 3d electrons, which sense the nuclear attraction efficiently [21]. The first principal calculation found that the Zn 3d electrons strongly interact with the O 2p electron in ZnO [47]. As the center energy of the green peak is similar than the band gap energy of ZnO (3.2eV), the visible emission can not be ascribed to the direct recombination of a conduction electron in the Zn 3d band and a hole in the O 2p valence band. Therefore, green emission must be related to the local level in band gap [21]. Infact the green emission results from the radiative recombination of photogenerated hole with an electron occupying the oxygen vacancy [44].

It has also been reported that there is difference in the spectra of ZnO of same size prepared by different methods [21, 45]. In our study also we have found that although size of nanoparticles remains unaltered, the process of annealing is giving rise to some changes in the emission spectra.

Researches indicate that the surface passivation via surfactant and polymer capping is an effective method to quench the defect-related visible photoluminescence (PL) from nanosized ZnO [48-50]. The effect of method of preparation and surface passivation itself is an indication that the green emission is due to surface states [48]. For the uncapped ZnO nanoparticles, as in our case, there exist abundant surface defects. Therefore, the absence of 551nm, 453nm peaks and appearance of 433.5nm peak in annealed ZnO nanoparticles as compared to those of unannealed particles, indicates the absence or presence of certain defect/surface states in both materials. As a result of this change in band structure, channels of visible emissions differ in unannealed and annealed nanoparticles.

It is earlier reported that ZnO nanopowders and thin films show green luminescence after they are annealed in oxygen, nitrogen or air [51]. In agreement with this finding, our results also demonstrate the green emission for annealed nanoparticles. Furthermore, it is revealed by our fluorescence studies that the green emission spread and other visible emissions are altered by annealing treatment.

B. Study of Ni doped ZnO nanostructures
The results and discussion pertaining to Ni:ZnO NS in the following sections.

B1. XRD analysis
The X-ray diffractograms of 1-10% Ni doped ZnO nanostructured powders are presented in figure 5.

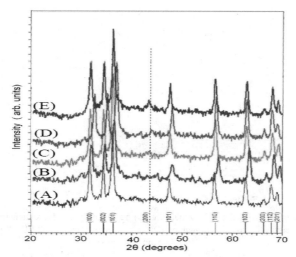

Figure 5. θ-2θ scans of ZnO nanostructured material doped with (A) 1%, (B) 3%, (C) 5%, (D) 7% and (E) 10% Ni concentrations respectively. The characteristic miller indices for hexagonal ZnO are also indicated for reference.

In all the diffractograms, the prominent X-ray diffraction lines at 31.8°, 34.6°, 36.3°, 47.6°, 56.7°, 62.9°, 66.4°, 68.0°, 69.1° could be indexed as (100), (002), (101), (102), (110), (103), (200), (112), (201), (004) and (202) respectively and correspond to reflections from hexagonal wurtzite phase of ZnO [34]. For 7% Ni:ZnO NS, an additional peak at 43.3° is also observed (as shown by dashed line) probably due to presence of NiO in the unreacted form in the lattice [35]. This peak gets pronounced for 10% Ni:ZnO NS [35]. For all the Ni:ZnO NS, the calculated lattice constants are almost same and were determined as a = 3.24Å, c = 5.21Å and c/a ratio is 1.579 which is similar to the theoretical value of 1.633. The average crystallite sizes of Ni:ZnO NS estimated by Scherrer's relation is in the range of 9-14nm, except for the 10% Ni:ZnO NS, for which it is 28nm [34]. This large change in the crystallite size prompted us to perform the microscopic characterizations of nanostructures.

B2. TEM, EDAX and SEM analysis
    In order to study the shape, size and particle distribution, we analyzed the TEM and SEM data for 1-10% Ni:ZnO NS, as shown in figures 6 and 7.

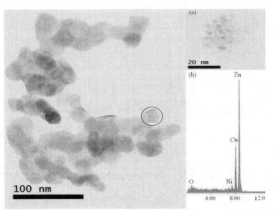

Figure 6. TEM image of 1% doped Ni:ZnO. Inset (a): resolved image of a particle (encircled in the TEM image) depicts the constituent nanoparticles of ~ 2nm size. Inset (b): EDAX pattern of 1% Ni:ZnO nanostructures.

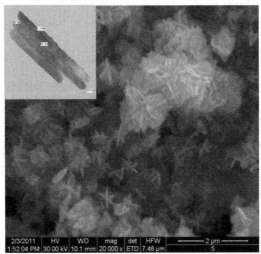

Figure 7. SEM image of 10% doped Ni:ZnO showing the formation of nanorods. The TEM image acquired for the same sample is also shown in the inset.

From the image shown in figure 6, the average particle size of 20 nm for 1% Ni:ZnO nanoparticles is confirmed. It is further noticed through the resolved image in the inset (a) that ZnO nanoparticles are indeed made up of much smaller sized nanoparticles of approximately 2nm size or even less. The similar observations recorded for 3%, 5% and 7% Ni:ZnO NS. Moreover, appearance of EDAX peaks of Zn, Ni and O, as seen in figure 6 (b), ascertains the formation of Ni:ZnO NS. The Cu peak in EDAX pattern is due to carbon coated copper grid used for mounting Ni:ZnO NS samples for

TEM/EDAX characterizations. No presence of nanoparticles could be detected in the SEM image of 10% Ni:ZnO (figure 7), however bunches of nanorods were vividly observed. It is of interest to note here that as the doping concentration of Ni increases to 7%, nanorods formation initiates along with nanoparticles of Ni:ZnO and for 10% Ni doping, growth of 215-264 nm long and 21-27 nm thick nanorods is predominately observed. The Ni:ZnO nanoparticles observed upto 5% Ni doping are in agreement with XRD analysis which suggest that wurtzite crystal structure of ZnO remain unaffected till 5% Ni substitution. However, when doping is further increased to 7% and 10%, the NiO (200) XRD peak is clearly seen. This is due to the fact that doping of Ni for 7% concentration or more is not completely substituted inside ZnO crystal. The excess of Ni doping assists in anisotropic growth of ZnO resulting in nanorod formation and the residual NiO crystallites remains with Ni:ZnO nanostructures.

CONCLUSION

We synthesized undoped and Ni doped ZnO nanostructures adopting a chemical precipitation technique without using any capping agent. In this method, growth of nanostructures is controlled by their pH concentration. In the first segment of the report it is demonstrated that annealing of undoped nanoparticles assists in removing the carbonate impurities and moisture so as to obtain pure ZnO nanocrystals. Our analyses suggest that undoped ZnO nanoparticles are stable and the typical problems of Ostwald ripening and agglomeration are not observed for the undoped nanoparticles prepared by this technique. The wurtzite phase of ZnO particles remains unaffected by annealing. The size estimated by XRD and TEM are in agreement. The blue shift observed in the UV-visible spectrum is the typical signature of size confinement in ZnO nanocrystals. The fluorescence study leads to interesting conclusion that although crystal phase and other structural parameters of ZnO nanoparticles are not influenced by annealing process, fluorescence emission is affected. This finding paves the way to the process of altering visible luminescence in undoped ZnO nanoparticles without varying their size or shape. The second segment of the report provides evidences of the crucial role of nickel doping in growth of Ni:ZnO nanoparticles and their structural evolution leading to the formation of Ni:ZnO nanorods.

REFERENCES
[1] A.P. Alivisatos, V.L. Colvin and A. N. Goldstein, Semiconductor Nanocrystals covalently bound to metal-surfaces with self-assembled monolayers, *J. Am. Chem. Soc.*, **114**, 5221-30 (1992).
[2] A.P. Alivisatos, Semiconductor clusters, nanocrystals and quantum dots, *Science*, **271**, 933-7 (1996).
[3] C. Nasr, S. Hotchandani, W.Y. Kim and P.V. Kamat, Photoelectrochemistry of composite semiconductor thin films photosensitization of SnO_2/CdS coupled nanocrystallites with a ruthenium polypyridyl complex, *J. Phys. Chem.*, **101**, 7480-87 (1997).
[4] Y.D. Li., X.L. Li, Z.X. Deng, B.C. Zhou, S.S. Fan, J.W. Wang and X.M. Sun, From inorganic-surfactant mesostructrue to metallic tungsten nanowire, *Angew. Chem. Int. Ed.*, **41**, 333-5 (2002).
[5] M.A. El-Sayed, Small Is Different: Shape-, Size-, and Composition-Dependent Properties of Some Colloidal Semiconductor Nanocrystals, *Acc. Chem. Res.*, **37**, 326-33 (2004).
[6] T. Bashe, F. Koberling and A. Mews, Oxygen-induced blinking of single CdSe nanocrystals, *Adv. Mater.* **13**, 672-6 (2001).
[7] W.U Huynh, J.J. Dittmer and A.P. Alivisatos, Hybrid nanorod-polymer solar cells, *Science* **295**, 2425-27 (2002).
[8] J.A. Ascencio, P. Santiago, L. Rendon and U. Pal, Structural basis of CdS nanorods:Synthesis and HRTEM characterization, *Appl. Phys. A*, **78**, 5-7 (2003).
[9] D. Yu, C.J. Wang and G.S. Philipe, n-type conducting CdSe nanocrystal solids, *Science,* **300**, 1277-80 (2003).

[10] J.P. Li, Y. Xu, D. Wu and Y.H. Sun, Hydrothermal synthesis of novel sandwich-like structured ZnS/octylamine hybrid nanosheets, *Solid State Commun.*, **130**, 619-22 (2004a).

[11] G. P. Mitchell, C.A. Mirkin and R.L. Letsinger, The programmed assembly of DNA functionalized quantum dots, *J. Am. Chem. Soc.*, **121**, 8122-23 (1999).

[12] W.U. Huynh, X.G. Peng and A.P. Alivisatos, CdSe nanocrystal rods/poly(3-hexylthiophene) composite photovoltaic devices, *Adv. Mater.*, **11**, 923-7 (1999).

[13] Z.R. Dai, Z.W. Pan and Z.L.Wang, Novel nanostructures of functional oxides synthesized by thermal evaporation, *Adv. Funct. Mater.*, **13**, 9-24 (2003).

[14] S. Saito, M. Miyayama, K. Koumoto, Gas sensing characteristics of porous ZnO and Pt/ZnO ceramics, *J Am. Ceram. Soc.*, **68**, 40-3 (1985).

[15] S.J. Pearton, D.P. Norton, K. Ip, Y.W. He and T. Steiner, Recent progress in processing and properties of ZnO, *Prog. Mater. Sci.* **50**, 293-340 (2005).

[16] B.S. Jeon, J.S. Yoo and J.D. Lee, Electrophoretic deposition of ZnO:Zn phosphor for field emission display applications, *J. Electrochem. Soc.*, **143**, 3923-29(1996).

[17] P.L. Hower and T.K. Gupta, A barrier model for ZnO varistors, *J. Appl. Phys.* **50**, 4847-55 (1979).

[18] W.G. Morris, Physical properties of the electrical barriers in varistors *J. Vac. Sci. Technol.*, **13**, 926-32 (1976).

[19] Z.L. Wang, Zinc oxide nanostructures: growth, properties and applications, *J. Phys.: Conds. Matter.* **16**, R829-R858 (2004).

[20] Z.L. Wang and J. Song, "Piezoelectric nanogenerators based on zinc oxide nanowire arrays *Science* **312**, 242-6 (2006).

[21] L. Irimpan, V. P. N. Nampoori, P. Radhakrishnan, A. Deepthy and B. Krishnan, Size dependent fluorescence spectroscopy of nanocolloids of ZnO, *J. Appl. Phys.*, **102**, 063524-30 (2007).

[22] L. Brus, Electronic wave functions in semiconductor clusters: experiment and theory, *J. Phys. Chem.*, **90**, 2555-60 (1986).

[23] Y. Yin and A. P. Alivisatos, Colloidal nanocrystal synthesis and the organic-inorganic interface, *Nature* **437**, 664-670 (2005).

[24] Z. Fan and J. G. Lu, Z. Fan and J.G. Lu, Zinc oxide nanostructures: Synthesis and properties, *J. Nanosci. and Nanotech,* **5**, 1561-73 (2005).

[25] J. Antony, X.B. Chen, J. Morrison, L. Bergman and Y. Qiang, ZnO nanoclusters: Synthesis and photoluminescence *Appl. Phys. Lett.* **87**, 241917-20 (2005).

[26] K.S. Choi, H.C. Lichtenegger and G.D. Stucky, Electrochemical synthesis of nanostructured ZnO films utilizing self-assembly of surfactant molecules at solid-liquid interfaces, *J. Am. Chem. Soc.* **124**, 12402-03 (2002).

[27] S. Maensiri, P. Laokul and V. Promarak, Synthesis and optical properties of nanocrystalline ZnO powders by a simple method using zinc acetate dihydrate and poly(vinylpyrrolidone)*J. Cryst. Growt,* **289**, 102-106 (2006).

[28] C. Wu, X. Qiao, J. Chen, H. Wang, F. Tan and S. Li, A novel chemical route to prepare ZnO nanoparticles, *Mater. Lett.,* **60**,1828-32 (2006).

[29] W.J. Li, E.W. Shi, W.Z. Zhong and Z.W. Yin, Growth mechanism and growth habit of oxide crystals, *J. Cryst. Growth,* **203**, 186-96 (1999).

[30] K.H. Tam, C.K. Cheung, Y.H. Leung, A.B. Djurišić, C.C. Ling, C.D. Beling, S. Fung, W.M. Kwok, W.K. Chan, D.L. Phillips, L. Ding, and W. K. Ge, Defects in ZnO nanorods prepared by a hydrothermal method, *J. Phys. Chem. B* **110**, 20865-71 (2006).

[31] C. Wu, X. Qiao, J. Chen, H. Wang, F. Tan and S. Li, A novel chemical route to prepare ZnO nanoparticles, *Mater. Lett.* **60**, 1828-32 (2006).

[32] Y. Hu and H.J. Chen, Preparation and characterisation of nanocrystalline ZnO particles from a hydrothermal process, *J. Nanopart. Res.* **10**, 401-07 (2008).

[33] R. Elilarassi and G. Chandrasekaran, Synthesis and optical properties of Ni-doped zinc oxide nanoparticles for optoelectronic applications, *Optoelectronic Letters*, **6**, 0006-10 (2010)

[34] N. Goswami and D.K. Sharma, Structural and optical properties of unannealed and annealed ZnO nanoparticles prepared by a chemical precipitation technique, *Physica E*, **42**, 1675-82 (2010)

[35] G. Peia, C. Xia, S. Caoc, J. Zhanga, F. Wua and J. Xua, Synthesis and magnetic properties of Ni-doped zinc oxide powders, *J. Magn. Magn. Mater.*, **302**, 340-2 (2006)

[36] K. Lepková, J. Clohessy and V.J. Cunnane, The pH-controlled synthesis of a gold nanoparticle/polymer matrix via electrodeposition at a liquid–liquid interface, *J. Phys.: Condens. Matter.* **19**, 375106-18 (2007).

[37] S. Steinfel, A.V. Gleich and U. Petschow, *Nanotechnologies Hazards and Resource Efficiency*, Springer Verlag, London (2007).

[38] D.K. Bhat, Facile Synthesis of ZnO Nanorods by Microwave Irradiation of Zinc–Hydrazine Hydrate Complex, *Nanoscale Res Lett.* **3**, 31-35 (2008).

[39] B. D. Cullity, *Elements of X-Ray Diffraction*, Addison-Wesley Publishing Company Inc. U.S.A (1978).

[40] L.V. Azaroff, *Elements of X-Ray Crystallography*, McGraw-Hill, U.S.A. (1968).

[41] N. Goswami and P. Sen, Photoluminescent properties of ZnS nanoparticles prepared by electro-explosion of Zn Wires, *Journal of Nanoparticle Research*, **9**, 513-17 (2007).

[42] G. Cao, *Nanostructures and Nanomaterals: Synthesis, Properties and Applications*, Imperial College Press, London (2006).

[43] R.M. Nyffenegger, B. Craft, M. Shaaban, S. Gorer, G. Erley and R.M. Penner, A hybrid electrochemical/chemical synthesis of zinc oxide nanoparticles and optically intrinsic thin films, *Chem. Mater.* **10**, 1120-29 (1998).

[44] K. Vanheusden, W.L. Warren, C.H. Seager, D.R. Tallant, J.A. Voigt and B.E. Gnade, Mechanishs behind green photoluminescence in ZnO phosphor powders, *J. Appl. Phys.* **79**, 7983-91 (1996).

[45] O. Madelung, *Semiconductors: Data Handbook*, Springer; 3rd edition (2003).

[46] L.V. Azaroff, *Introduction to Solids*, p.371-372, McGraw-Hill, New York (1960).

[47] P. Schroer, P. Kruger and J. Pollmann, First-principles calculation of the electronic structure of the wurtzite semiconductors ZnO and ZnS, *Phys. Rev. B* **47**, 697-80 (1993).

[48] D. Li, Y. H. Leung, A. B. Djurisic, Z. T. Liu, M. H. Xie, S. L. Shi, S. J. Xu and W. K. Chan, Different Origins of visible luminescence in ZnO nanostructures fabricated by the chemical and evaporation methods, *Appl. Phys. Lett.* **85**, 1601-04 (2004).

[49] L. Guo, S. Yang, C. Yang, P. Yu, J. Wang, W. Ge and G. K. L. Wong, Highly monodisperse polymer-capped ZnO nanoparticles: Preparation and optical properties, *Appl. Phys. Lett.* **76**, 2901-04 (2000).

[50] V.A. Fonoberov and A. Balandin, Origin of ultraviolet photoluminescence in ZnO quantum dots: Confined excitons versus surface-bound impurity exciton complexes, *Appl. Phys. Lett.* **85**, 5971-74 (2004).

[51] J.Z. Wang, G. T. Du, Y. T. Zhang, B.J. Zhao, X. T. Yang and D. L. Liu, Luminescence properties of ZnO films annealed in growth ambient and oxygen, *J. Cryst. Growth*, **263**, 269-272 (2004).

CHEMICAL VAPOR DEPOSITION GROWTH OF GRAPHENE ENCAPSULATED PALLADIUM NANOPARTICLES

Junchi Wu, Nitin Chopra*
Department of Metallurgical and Materials Engineering,
Center for Materials for Information Technology (MINT)
University of Alabama
Tuscaloosa, AL-35487, USA
*Email: nchopra@eng.ua.edu
Tel: 205-348-4153
Fax: 205-348-2164

ABSTRACT

This paper reports synthesis of palladium nanoparticles with emphasis on structural and morphological evolution. This was investigated by varying surfactant concentration, metal salt concentration, temperature, and growth duration for the wet-chemical synthesis approach employed to grow palladium nanoparticles. It was observed that cubic palladium nanoparticles were formed by adding shape controlling additives (potassium chloride and potassium bromide). The produced cubic palladium nanoparticles were oxidized in an oxygen plasma treatment process. Transmission electron microscopy and X-ray photoelectron spectroscopy were utilized to characterize morphology, crystal structure, and chemical states of palladium nanoparticles as well as surface oxidized nanoparticles. The oxidized cubic palladium nanoparticles were further used as catalyst for the growth of graphene shells (~ 2- 5 nm) in a chemical vapor deposition process. The synthesized graphene encapsulated palladium nanoparticles were studied using electron microscopy and Raman spectroscopy.

INTRODUCTION

Over the past two decades, metal nanoparticles have received significant attention due to their unique size-dependent electronic, physical, and chemical properties.[1-3] Palladium nanoparticles (PdNPs) have applications in catalysis, hydrogen storage, and chemical sensing.[4-8] However, one of the biggest challenges is to prevent the aggregation of PdNPs.[9] For example, utilization of PdNPs in catalytic converters is limited due to their high temperature sintering, resulting in the reduction of surface area-to-volume ratio, affecting the catalytic activity of these nanoparticles.[10] Similarly, aggregation due to surface charging and electrostatics hinders longevity of chemical sensors employing PdNPs.[11,12] Thus, it is critical to encapsulate PdNPs in a robust cage or a shell that can survive harsh environments in solution or gas phase.

Carbon nanotubes (CNTs) have been filled with various nanoparticles including PdNPs.[13,14] Most of the CNT filling techniques were post-growth methods involving acid-based opening of CNT cores and filling through capillary action or wet-chemistry. Such an approach also results in the contamination of final heterostructure and adds to the cost of processing. Moreover, filling nanotubular structures might lead to incomplete filling of the former and lead to entrapment of fluids in these spaces due to the 1-D geometry. In this regard, encapsulating PdNPs in a graphene or carbon onion shell is promising. This hybrid nanoparticle configuration (core/shell) inherits properties of both components and the rich surface chemistry of graphene shell. Carbon encapsulated metal nanoparticles with core metal as Sn, Pt, Au, and Ag have been reported earlier.[15-18] The synthesis approaches to obtain carbon shells around these nanoparticles include arc-discharge, ion-beam sputtering, high-temperature electron irradiation, or pyrolysis of organic precursors.[19-22] In addition, "wormlike" palladium-carbon has been synthesized by electric arc treated PdO/graphite mixture.[23] Vajtai and co-workers fabricated carbon nanotube (CNT) network from palladium patterned substrates.[24] Formation of palladium/carbon core/shell structures through hydrothermal process at 200 °C has also been reported,[25] but the carbon shell was disordered and thick (~ 5 - 45 nm with varying reaction time). Despite the formation of the abovementioned palladium-carbon heterostructures, thin graphene encapsulated PdNPs are not reported. Recently, a thermal CVD approach for synthesizing graphene encapsulated AuNPs has been reported.[26,27] This method has proved to be efficient in producing large silicon substrate area covered with graphene encapsulated AuNPs. In order to explore versatility of our CVD approach, we report the synthesis of graphene encapsulated PdNPs and outline their growth mechanism here. The first step towards this goal is to control the morphology and structure of PdNPs using a simple wet-chemistry approach. A key to successful growth of graphene encapsulated PdNPs is to utilize surface-oxidize PdNPs as a catalyst, thus plasma oxidation of PdNPs is also reported here.

EXPERIMENTAL

Synthesis of palladium nanoparticles

PdNPs were synthesized by ethanol reduction approach.[28] Approximately, 0.027 mmol PdCl$_2$ was dissolved in a mixture of 10 mL ethanol and 10 mL DI water. To this solution, ~ 1.044 mmol polyvinylpyrrolidone (PVP) was added and the solution was allowed to reflux at 100 °C for 3 h in N$_2$ atmosphere. A dark brown precipitate was observed after the reaction, washed with copious amounts of DI water and acetone. The cleaned precipitate was finally dispersed in DI water. To examine the influence of different growth parameters on the size and morphology of PdNPs, systematic studies (Table I) were performed by changing surfactant concentration (0.013 – 0.1 mM), metal salt concentration (0.001 – 0.025 mM), temperature (100 – 150 °C), or reaction duration (1 – 10 h). Only one variable was changed while keeping all the other experimental conditions fixed for each particular experiment. To control the shape of PdNPs, potassium bromide (KBr) and potassium chloride (KCl) were added as shape-control additives. An aqueous solution (~ 8.0 mL), containing ~ 0.945 mmol PVP, ~ 0.341 mmol L-ascorbic acid (AA), ~ 5.042 mmol KBr, ~ 2.482 mmol KCl were placed in a volumetric flask, refluxed and pre-heated in oil bath under magnetic stirring at 80 °C for 10 min. Subsequently, ~ 3.0 mL of an aqueous solution containing ~ 0.194 mmol sodium tetrachloropalladate (II) hexahydrate (Na$_2$PdCl$_4$·xH$_2$O (x≈3)) was added using a pipette. The solution was stirred and refluxed at 80 °C for 3 h. The dark brown solution was centrifuged, washed to remove excess PVP, and re-dispersed in 11 mL DI water.

Table I. Systematic study for synthesis of PdNPs. The shaded region in the table denotes the varying parameter, and baseline parameters (sample 3, bold and italicized fonts).

Sample number	Metal salt	Reducing agent	PVP amount/mM	Metal salt/mM	T/°C	Time/h	KBr/mM	KCl/mM
1	PdCl$_2$	Ethanol	0.013	0.001	100	3	0	0
2	PdCl$_2$	Ethanol	0.026	0.001	100	3	0	0
3	*PdCl$_2$*	*Ethanol*	*0.052*	*0.001*	*100*	*3*	*0*	*0*
4	PdCl$_2$	Ethanol	0.1	0.001	100	3	0	0
3	*PdCl$_2$*	*Ethanol*	*0.052*	*0.001*	*100*	*3*	*0*	*0*
5	PdCl$_2$	Ethanol	0.052	0.003	100	3	0	0
6	PdCl$_2$	Ethanol	0.052	0.013	100	3	0	0
7	PdCl$_2$	Ethanol	0.052	0.025	100	3	0	0
3	*PdCl$_2$*	*Ethanol*	*0.052*	*0.001*	*100*	*3*	*0*	*0*

8	PdCl$_2$	Ethanol	0.052	0.001	125	3	0	0
9	PdCl$_2$	Ethanol	0.052	0.001	150	3	0	0
10	PdCl$_2$	Ethanol	0.052	0.001	100	1	0	0
3	*PdCl$_2$*	*Ethanol*	*0.052*	*0.001*	*100*	*3*	*0*	*0*
11	PdCl$_2$	Ethanol	0.052	0.001	100	6	0	0
12	PdCl$_2$	Ethanol	0.052	0.001	100	10	0	0
13	Na$_2$PdCl$_4$	AA	0.086	0.018	80	3	0.458	0.225

Plasma oxidation of nanoparticles

As-synthesized cubic PdNPs (Table I, sample 13) were dispersed on piranha cleaned silicon wafer by drop-casting method and dried in vacuum oven. This was followed by plasma oxidation process. The oxidation was performed in Model 1021 plasma cleaner at 600 W and ~ 36 mTorr of mixture gas (25% O$_2$ and 75% Ar) pressure. The dispersed PdNPs were plasma oxidized for 30 min to result in surface oxide.

Growth of graphene encapsulated PdNPs

Surface oxidized PdNPs were utilized as a catalyst for the growth of graphene shells in a CVD process.[26,27,29-31] Silicon wafer decorated with oxidized PdNPs was placed in the center of a quartz tube equipped with precursors and gas lines for Ar/H$_2$ flow. Xylene was injected through a syringe injector at the rate of 45 mL/h for about 2 min and subsequently transported into the reaction zone (~ 675 °C) inside the quartz tubes furnace. The xylene flow rate was reduced to 10 mL/h when H$_2$ in Ar (Ar/H$_2$=1.8 SLM: 0.2 SLM 10% v/v) carrier gas was introduced in the CVD reactor, where H$_2$ acted as an oxygen scavenger. The CVD reaction was continued for 1 h after which H$_2$ and xylene flow was discontinued and furnace was cooled down under Ar flow.

Characterization

Tecnai F-20 was used to collect Transmission Electron Microscopy (TEM) images at 200 kV. TEM samples were prepared by dispersing as-prepared samples on lacey carbon TEM copper grids. The average nanoparticle size was measured from TEM images, where 200 nanoparticles were counted and measured per sample. Diameter was measured for spherical nanoparticles, average side length for triangular, and for nanoparticles with other shapes, diagonal length average was taken. All the measurements were done using Adobe Photoshop Software. X-ray photoelectron spectra (XPS) were gathered by Kratos Axis 165 with Aluminum mono-gun. The analysis spot was set as "Slot" with >20 μm aperture and 19.05 mm iris setting. XPS was used to characterize elements' chemical states. Raman spectra were collected using Bruker Senterra system (Bruker Optics Inc. Woodlands, TX) equipped with 785 nm laser source at 10-25 mW laser powers and 100X objective. The integral time and co-additions were set as 100 seconds and 2, respectively.

RESULTS AND DISCUSSION

Synthesis and characterization of PdNPs

The PdNPs were synthesized in the described wet-chemical process and the result summary of the systematic growth studies are indicated in table I. The synthesized PdNPs were characterized using TEM as shown in Figures 1-5. The trends in average nanoparticle sizes corresponding to various growth conditions (in the order shown in Table I) are also shown in corresponding figures (Figure 1-5). A decrease of nanoparticle size was observed when higher amount of PVP was utilized (~ 6.3 to 4.0 nm, Figure 1E). PVP is a widely used stabilizer for nanoparticle synthesis.[32] Increasing PVP concentration resulted in tight packing of stabilizer micelle around the nucleated PdNPs and this hindered the diffusion of palladium ions into the micelle. This is the reason for decrement of the PdNPs size with increasing PVP concentration. It was observed that larger size was obtained as the metal salt concentration was increased (~ 3.6 to 8.4 nm, Figure 2D). This is attributed to a higher supersaturation of metal salt that enhances both nucleation and growth rate.[33] Similar has been observed for the wet-chemical growth of Ni nanoparticles by employing metal salt, stabilizers, and complexing agent.[34] As the growth temperature increased from 100 °C to 150 °C, PdNPs size slightly increased from 3.6 ±1.2 nm to 5.3 ± 3.2 nm (Figure 3C). This indicates that temperature does not have significant effect on the nanoparticle size. Moreover, as the growth duration increases, the average size of PdNPs increases (~ 4.7 to 8.1 nm, Figure 4D). This is due to the fact that longer reaction duration allows for growth of nanoparticles as long as the reactants are not consumed. Table II shows different shapes and lattice spacing present in samples consistent with the growth conditions given in table I. The authors believe that this also represents true data set as it is impossible to obtain 100% spherical nanoparticles for any growth method. Even though minor but this sample heterogeneity is very critical for further utilization of PdNPs for their applications. It should be noted that most of the growth conditions resulted in greater than 90% spherical nanoparticles. However, sample 7 that corresponds to the highest metal salt to PVP ratio resulted in lowest percentage of spherical PdNPs. The former ratio is a critical factor for PdNPs growth and higher this ratio is, the greater the possibility of less tightly packed micelles around growing PdNPs.[34] This could lead to more anisotropic growth of PdNPs further allowing for the formation of different shapes. Moreover, if the shape inducing reactants are used (KBr and KCl) then no spherical PdNPs are formed as shown by results for sample 13 (Table II, Figure 5). The average size of the cubic nanoparticles was observed to be ~ 17.4±2.6 nm (Figure 5). In regard to crystal planes shown, lowest energy (111) planes are dominantly present in all the samples (Table II).[33] On the other hand, using halide species such as Br⁻ and Cl⁻ during the synthesis allowed for developing (200) planes (Figure 5). This is attributed to the affinity of Br⁻ and Cl⁻ for {100} facets of PdNPs.[35]

Table II. Result summary for systematic study indicating average size, shape distribution, lattice spacing, and the corresponding planes for the as-prepared PdNPs according to the table I.

Sample number	Average size (nm)	Shape distribution	Lattice spacing (nm) and the corresponding plane
1	6.3±2.7	Hexagon: 4.2%, Triangle: 1.3%, Spherical: 94.5%	0.235±0.004 (111)
2	6.2±3.8	Hexagon: 4.2%, Rhombus: 0.8%, Square: 0.4%, Triangle: 1.7%, Spherical: 92.9%	0.232±0.008 (111)
3	3.6±1.2	Hexagon: 2.8%, Triangle: 0.9%, Pentagon: 0.9%, Square: 0.9%, Spherical: 94.5%	0.198 (200) 0.234±0.001 (111)
4	4.0±1.8	Hexagon: 3.4%, Pentagon: 0.5%, Rhombus: 0.2%, Square: 0.7%, Triangle: 0.2%, Spherical: 95%	0.201 (200) 0.232±0.003 (111)
5	4.5±2.1	Hexagon: 1.4%, Rhombus: 0.9%, Square: 2%, Triangle: 3.7%, Spherical: 92%	0.234±0.011 (111) 0.196 (200)
6	4.5±1.7	Hexagon: 0.2%, Pentagon: 0.4%, Rhombus: 0.6%, Square: 0.6%, Triangle: 0.8%, Spherical: 97.4%	0.204 (200) 0.230±0.005 (111)
7	8.4±2.9	Hexagon: 7.6%, Rod: 0.8%, Rhombus: 3%, Square:	0.230±0.003 (111)

		2.3%, Triangle: 3%, Spherical: 83.3%	
8	5.2±3.7	Hexagon: 3.6%, Pentagon: 0.65%, Rhombus: 2.8%, Square: 0.43%, Triangle: 1.9%, spherical: 90.62%	0.198 (200) 0.229±0.002 (111)
9	5.3±3.2	Hexagon: 1.8%, Pentagon: 0.3%, Rhombus: 0.6%, Square: 0.9%, Triangle: 2.4%, Spherical: 94.0%	0.204(200) 0.229±0.003 (111)
10	4.7±2.1	Hexagon: 0.8%, Pentagon: 0.8%, Square: 1.3%, Triangle: 3.2%, Spherical: 93.8%	0.203±0.002 (200) 0.229±0.005 (111)
11	6.8±3.8	Hexagon: 2.4%, Pentagon: 0.3%, Rod: 0.3%, Rhombus: 1.5%, Square: 1.2%, Triangle: 1.5%, Spherical: 92.8%	0.229±0.001 (111) 0.199 (200)
12	8.1±4.0	Hexagon: 4.4%, Pentagon: 0.4%, Rod: 0.4%, Rhombus: 0.3%, Square: 1.6%, Triangle: 1.5%, Spherical: 91.4%	0.230±0.007 (111) 0.199 (200)
13	17.4±2.6	Cubic 89.6%; Triangle: 8.3%; Pentagon: 2.1%	0.196±0.002 (200)

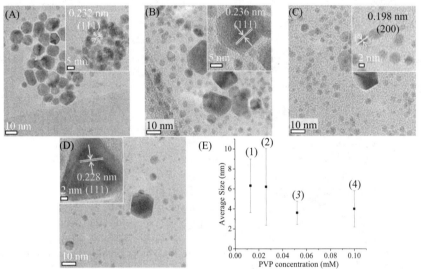

Figure 1. TEM images of PdNPs (table I, sample 1-4) synthesized using different PVP concentrations: (A) 0.013 mM, (B) 0.026 mM, (C) 0.052 mM, (D) 0.1 mM, and (E) size distribution of PdNPs. Note: The number in the parentheses indicates the sample number. Sample 3 is the baseline sample.

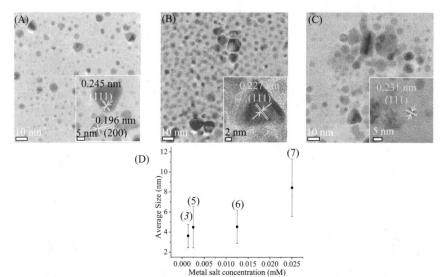

Figure 2. TEM images of PdNPs (table I, sample 5-7) synthesized using different metal salt (PdCl$_2$) concentration: (A) 0.003 mM, (B) 0.013 mM, (C) 0.025 mM, and (D) size distribution of PdNPs. Note: The number in the parentheses indicates the sample number. Sample 3 is the baseline sample.

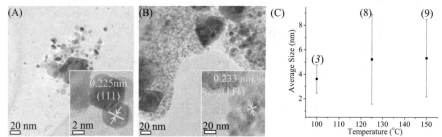

Figure 3. TEM images of PdNPs (table I, sample 8-9) synthesized at different temperatures: (A) 125 °C, (B) 150 °C, and (C) size distribution of PdNPs. Note: The number in the parentheses indicates the sample number. Sample 3 is the baseline sample.

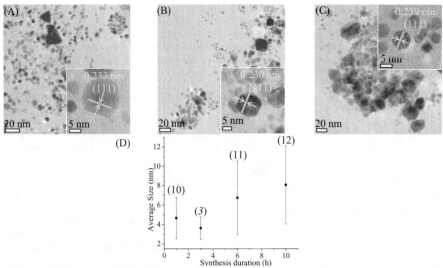

Figure 4. TEM images of PdNPs (table I, sample 10-12) synthesized for different growth duration: (A) 1 h, (B) 6 h, (C) 10 h, and (D) size distribution of PdNPs. Note: The number in the parentheses indicates the sample number. Sample 3 is the baseline sample.

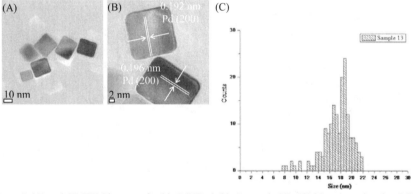

Figure 5. (A) and (B) TEM images of cubic PdNPs (table I, sample 13), (C) histogram showing PdNPs size distribution.

Plasma oxidation of palladium nanoparticles

As synthesized cubic PdNPs (Table I, sample 13) were selected for plasma oxidation process. XPS was utilized to determine formation of palladium oxide. Figure 6A shows an overlayed XPS survey scans with binding energy ranging from 1000 to 0 eV and shows chemical composition of nanoparticles before and after plasma oxidation for 30 min. Elements Pd, Na, C, and Si presented in spectra were from PdNPs, metal salt, PVP, and silicon substrate, respectively. In regard to O peaks, apart from palladium oxide, the absorbed organic molecules and surface SiO_2 from substrate can also attribute to the peak intensity. Thus the formation of palladium oxide could only be detected by detailed analysis of Pd XPS peaks. Figure 6B and C shows detailed deconvolution of Pd 3d peaks and their binding energies.

It was observed that as-prepared PdNPs resulted in Pd peaks consistent with Pd $3d_{5/2}$ and $3d_{3/2}$ at ~335.16 eV and 340.44 eV, respectively. PdO and PdO_2 peaks were observed at 336.50 eV ($3d_{5/2}$, PdO), 341.81 eV ($3d_{3/2}$, PdO), 338.12 eV ($3d_{5/2}$, PdO_2), and 343.09 eV ($3d_{3/2}$, PdO_2), respectively (Figure 6B and C).[36,37] The formation of palladium oxide before plasma could be attributed to the air oxidation of PdNPs during synthesis, especially on the corners of cubic-shaped nanoparticles. In addition, dissolved air or oxygen species could also react with high surface-to-volume ratio PdNPs during the synthesis at synthetic temperature (80 °C).[38] After plasma oxidation for 30 min, the peaks corresponding to palladium oxide increased and ensured the surface oxidation of PdNPs. PdO was the major component as observed by large intensity peaks for the same (Figure 6B). This also suggests that PdO is more stable in the oxidized product. Overall, it was observed that ~ 85% (at.%) palladium oxide was formed as the plasma oxidation process was continued till 30 min. The mole fraction of PdO before oxidation is lower (~ 35%, at.%) than after plasma treated one. This was obtained by calculating the ratio of the area under the palladium oxide peak to total area under all the peaks in the XPS spectra.

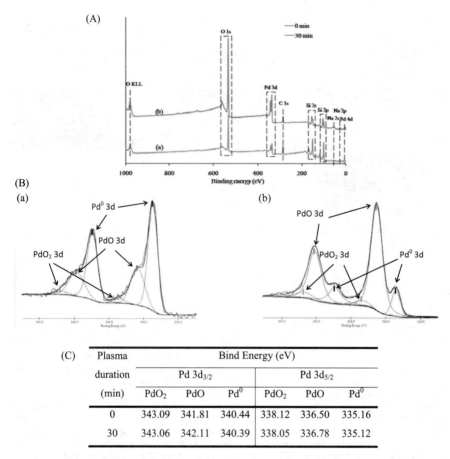

(C) Plasma	Bind Energy (eV)					
duration	Pd 3d$_{3/2}$			Pd 3d$_{5/2}$		
(min)	PdO$_2$	PdO	Pd^0	PdO$_2$	PdO	Pd^0
0	343.09	341.81	340.44	338.12	336.50	335.16
30	343.06	342.11	340.39	338.05	336.78	335.12

Figure 6. (A) XPS spectra corresponding to PdNPs (a) before plasma oxidation and after (b) 30 min of plasma oxidation duration, (B) deconvulated Pd peaks, (C) binding energy (eV) derived from XPS study for cubic PdNPs before and after plasma oxidation.

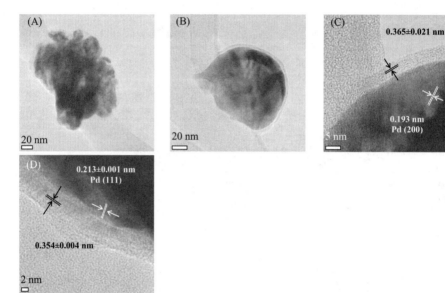

Figure 7. (A-D) TEM images of graphene encapsulated PdNPs on Si wafer.

TEM images for graphene encapsulated PdNPs (Figure 7) show their aggregation and surface migration during the CVD process. Clustering of nanoparticles took place, the size of PdNPs increased from ~ 20 nm to ~ 100 nm, and the shape of particles modified from cubic to near spherical. It should be noted that PdNPs were dispersed on silicon substrate by drop-casting, the high density of particles and the close distance could further lead to their aggregation. The authors expect to overcome this problem by chemically patterning particles on substrate with far away inter-particle distance and optimized graphene growth time. In regard to core PdNPs crystal planes, (111) and (200) planes are dominantly observed (Figure 7C and D). The observed graphene shells have a thickness around 2 to 5 nm, with lattice spacing of ~ 0.35 nm, well fitted to c-axis spacing of graphene (Figure 7C and D). Some amorphous carbon also formed during the graphene growth process, which was further confirmed by Raman spectroscopy (Figure 8) that clearly showed D (~ 1300 cm^{-1}) and G bands (~ 1580 cm^{-1}) corresponding to disordered and graphitic carbon content, respectively. We propose the following growth mechanism for the graphene encapsulated PdNPs. During CVD growth, xylene decomposed at the growth temperature. Thereafter, the surface palladium oxide, which is unstable at high temperature,[39] gets reduced and accepts electrons from incoming carbon feed. This results in PdNPs encapsulated in crystallized carbon (or graphene) shell.

In order to achieve enhanced chemical functionality in graphene encapsulated PdNPs, it will be critical to partially expose the surface of core PdNPs. Thus, the on-going research in the authors' laboratory is focused on selective etching of the graphene shells to create porosity that provides partial accessibility to the Pd core.[18] Electron beam irradiation, plasma, and chemical methods can help to partially open carbon/graphene shells.[40-42] This has been demonstrated by the authors in their earlier study on graphene encapsulated AuNPs, where plasma etching was used to create porous/etched graphene shells around AuNPs and exposing the surface of the core nanoparticle.[27]

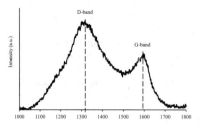

Figure 8. Raman spectrum of graphene encapsulated PdNPs showing the D and G band.

CONCLUSIONS

The influence of PVP concentration, metal salt concentration, temperature, and growth duration on the structure and morphology of PdNPs was studied. Increasing surfactant concentration resulted in the reduction in size of the PdNPs. High concentration of metal salt and longer growth duration enhanced the nucleation and growth process and led to increased average size of the nanoparticles. The shape of PdNPs can be well controlled by adding KBr and KCl to increase anisotropic growth, resulting in cubic PdNPs with {100} facets. Cubic PdNPs were then surface oxidized by plasma oxidation. XPS was utilized to understand the oxidation process. Palladium oxide is effectively formed during oxygen plasma treatment, in which PdO is more stable as indicated by the XPS results.

Graphene shell growth was performed on surface-oxidized cubic PdNPs in a CVD process. The oxidation of PdNPs is a critical step for graphene shell growth. As proposed here, the unstable surface oxide (at high growth temperatures) will reduce and allow for the formation of the graphene shell. The thickness of the graphene shell was observed to be ~ 2 - 5 nm with lattice spacing of ~ 0.35 nm indicating the formation of graphene structure. From Raman spectroscopy, D-band and G-band are distinguished after graphene growth. Future work will be focused on deriving experimental proof of the proposed growth mechanism. Such hybrid nanoparticle configuration holds strong potential for applications in energy, sensing, and catalysis.

ACKNOWLEDGMENTS

The work is supported by the National Science Foundation under Grant No. (0925445). The authors also thank the University of Alabama (MTE department and the Office of sponsored programs), MSE PhD program, and MINT center for the start-up funds and for financial aid for J.W. The authors also thank the Central Analytical Facility (CAF) for electron microscopy equipment and the financial support covering the instrument time, the MINT Center for providing infrastructure support such as clean room facility and various equipments, and Mr. Rich Martens, Mr. Johnny Goodwin, and Mr. Rob Holler for providing training on microscopy and surface analysis instruments. The authors also thank Dr. Shweta Kapoor for proof reading the manuscript.

REFERENCES

[1] N. Satoh, T. Nakashima, K. Kamikura, and K. Yamamoto, Quantum Size Effect in TiO$_2$ Nanoparticles Prepared by Finely Controlled Metal Assembly on Dendrimer templates, *Nat. Nanotechnol.*, **3** (2), 106-11(2008).

[2] H. Jiang, K. Moon, H. Dong, F. Hua, C. Wong, Size-dependent Melting Properties of Tin Nanoparticles, *Chem. Phys. Lett.*, **429** (4-6), 492-96 (2006).

[3] X. Chen, S. Mao, Titanium Dioxide Nanomaterials: Synthesis, Properties, Modifications, and Applications, *Chem. Rev.*, **107** (7), 2891-59 (2007).

[4] M. Kwon, N. Kim, C. Park, J. Lee, K. Kang, and J. Park, Palladium Nanoparticles Entrapped in Aluminum Hydroxide: Dual Catalyst for Alkene Hydrogenation and Aerobic Alcohol Oxidation, *Org. Lett.*, **7** (6), 1077-79 (2005).

[5] M. Gray, W. McCaffrey, Role of Chain Reactions and Olefin Formation in Cracking, Hydroconversion, and Coking of Petroleum and Bitumen Fractions, *Energ. Fuel.*, **16** (3), 756-66 (2002).

[6] M. Jin, H. Liu, H. Zhang, Z. Xie, J. Liu, and Y. Xia, Synthesis of Pd Nanocrystals Enclosed by {100} Facets and with Sizes < 10 nm for Application in CO Oxidation, *Nano Res.*, **4** (1), 83-91 (2011).

[7] S. J. Kishore, J. Nelson, Adair and P. Eklund, Hydrogen Storage in Spherical and Platelet Palladium Nanoparticles, *J. Alloys Compd.*, **389**,234-42 (2005).

[8] Y. Lu, J. Li, J. Han, H.-T. Ng, C. Binder, C. Partridge, M. Meyyappan, Room Temperature Methane Detection Using Palladium Loaded Single-Walled Carbon Nanotube Sensors, *Chem. Phys. Lett.*, **391**, 344-48 (2004).

[9] A. Biffis, M. Zecca, and M. Basato, Palladium Metal Catalysts in Heck C-C Coupling Reactions, *J. Mol. Catal. A-Chem.*, **173** (1-2), 249-74 (2001).

[10] J. Park, A. Forman, W. Tang, J. Cheng, Y. Hu, H. Lin, and E. McFarland, Highly Active and Sinter-Resistant Pd-Nanoparticle Catalysts Encapsulated in Silica, *Small*, **4** (10), 1694-97 (2008).

[11] W. Wongwiriyapan, Y. Okabayashi, S. Minami, K. Itabashi, T. Ueda, R. Shimazaki, T. Ito, K. Oura, S. Honda, H. Tabata, and M. Katayama, Hydrogen Sensing properties of Protective-Layer-Coated Single-Walled Carbon Nanotubes with Palladium Nanoparticle Decoration, Nanotechnology, **22**, 05501 (2011).

[12] O. Cernohorsky, K. Zdansky, J. Zavadil, P. Kacerovsky and K. Piksova, Palladium Nanoparticles on InP for Hydrogen Detection, Nanoscale Res. Lett., **6**, 410 (2011).

[13] M. Meyyappan, Carbon Nanotubes: Science and Applications, CRC Press, Boca Raton, FL (2005).

[14] D. Bera, S. Kuiry, M. McCutchen, A. Kruize, H. Heinrich, M. Meyyappan, and S. Seal, In-situ Synthesis of Palladium Nanoparticles-Filled Carbon Nanotubes Using Arc-discharge in Solution, *Chem. Phys. Lett.*, **386** (4-6), 364-68 (2004).

[15] S. Guo, J. Gong, P. Jiang, M. Wu, Y. Lu, and S. Yu, Biocompatible, Luminescent Silver@Phenol Formaldehyde Resin Core/Shell Nanospheres: Large-scale Synthesis and Application for in Vivo Bioimaging, *Adv. Funct. Mater.*, **18** (6), 872-79 (2008).

[16] Z. Wen, J. Liu, and J. Li, Core/shell Pt/C Nanoparticles Embedded in Mesoporous Carbon as a Methanol-Tolerant Cathode Catalyst in Direct Methanol Fuel Cells, *Adv. Mater.*, **20** (4), 743-47 (2008).

[17] Y. Yu, L. Gu, C. Wang, A. Dhanabalan, P. Aken, J. Maier, Encapsulation of Sn@Carbon Nanoparticles in Bamboo-like Hollow Carbon Nanofibers as an Anode Material in Lithium-Based Batteries, *Angew Chem Int Edit.*, **48** (35), 6485-89 (2009).

[18] T. Harada, S. Ikeda, F. Hashimoto, T. Sakata, K. Ikeue, T. Torimoto, and M. Matsumura, Catalytic Activity and Regeneration Property of a Pd Nanoparticle Encapsulated in a Hollow Porous Carbon Sphere for Aerobic Alcohol Oxidation, *Langmuir*, **26** (22), 17720-25 (2010).

[19] D. Ugarte, How to Fill or Empty a Graphitic Onion, *Chem. Phys. Lett.*, **209** (1-2), 99-03 (1993).

[20] T. Hayashi, S. Hirono, M. Tomita, and S. Umemura, Magnetic Thin Films of Cobalt Nanocrystals Encapsulated in Graphite-Like Carbon, *Nature*, **381** (6585), 772-74 (1996).

[21] E. Sutter, P. Sutter, and Y. Zhu, Assembly and Interaction of Au/C Core-Shell Nanostructures: In Situ Observation in the Transmission Electron Microscope, *Nano Lett.*, **5** (10), 2092-96 (2005).

[22] J. Kim, C. Kim, Y. Kim, and C. Yoon, Synthesis of Carbon-Encapsulated Gold Nanoparticles in Polyimide Matrix, *Colloid. Surface. A.*, **321** (1-3), 297-00 (2008).

[23] Y. Wang, Encapsulation of Palladium Crystallites in Carbon and the Formation of Wormlike, *J. Am. Chem. Soc.*, **116** (1), 397-98 (1994).

[24] R. Vajtai, K. Kordas, B. Wei, J. Bekesi, S. Leppavuori, T. George, and P. Ajayan, Carbon Nanotube Network Growth on Palladium Seeds, *Mat. Sci. Eng. C-Bio. S.*, **19** (1-2), 271-74 (2002)

[25] W. Kang, H. Li, Y. Yan, P. Xiao, L. Zhu, K. Tang, Y. Zhu, and Y. Qian, Worm-Like Palladium/Carbon Core-Shell Nanocomposites: One-Step Hydrothermal Reduction-Carbonization Synthesis and Electrocatalytic Activity, *J. Phys. Chem. C.*, **115** (14), 6250-56 (2011).

[26] N. Chopra, L. Bachas, and M. Knecht, Fabrication and Biofunctionalization of Carbon-Encapsulated Au Nanoparticles, *Chem. Mater.*, **21** (7), 1176-78 (2009).

[27] J. Wu, N. Chopra, Graphene Encapsulated Gold Nanoparticles and their Characterization, *Ceram. Trans.*, 223 (2010).

[28] O. Masala, R.Seshadri, Synthesis Routes for Large Volumes of Nanoparticles, *Ann. Rev. Mater. Res.*, **34**, 41-81 (2004).

[29] R. Andrews, D. Jacques, A. Rao, F. Derbyshire, D. Qian, X. Fan, E. Dickey, and J. Chen, Continuous Production of Aligned Carbon Nanotubes: a Step Closer to Commercial Realization, *Chem. Phys. Lett.*, **303** (5-6), 467-74 (1999).

[30] N. Chopra, P. Kichambare, R. Andrews, and B. Hinds, Control of multiwalled Carbon Nanotube Diameter by Selective Growth on the Exposed Edge of a Thin film Multilayer Structure, *Nano Lett.*, **2** (10), 1177-81 (2002).

[31] N. Chopra, B. Hinds, Catalytic Size Control of Multiwalled Carbon Nanotube Diameter in Xylene Chemical Vapor Deposition Process, *Inorg. Chim Acta.*, **357** (13), 3920-26 (2004).

[32] H. Wang, X. Qiao, J. Chen, X. Wang, and S. Ding, Mechanisms of PVP in the Preparation of Silver Nanoparticles, *Mater. Chem Phys.*, **94** (2-3), 449-53 (2005).

[33] G. Cao, Nanostructure and Nanomaterials, *Imperial College Press*, London, UK (2004).

[34] N. Chopra, L. Claypoole (NSF-REU), L.Bachas, Morphological Control of Ni/NiO Core/Shell Nanoparticles and Production of Hollow NiO Nanostructures, *J. Nanopart. Res.*, **12**, 2883-93 (2010).

[35] Y. Xiong, H. Cai, B. Wiley, J. Wang, M. Kim, and Y. Xia, Synthesis and Mechanistic Study of Palladium Nanobars and Nanorods, *J. Am. Chem. Soc.*, **129** (12), 3665-75. (2007)

[36] K. Kim, A. Gossmann, and N. Winograd, X-ray Photoelectron Spectroscopic Studies of Palladium Oxides and the Palladium-Oxygen Electrode, *Anal. Chem.*, **46** (2), 197-00 (1974).

[37] C. Wagner, Handbook of XPS, *Physical Electronics, Inc.*, (1979).

[38] R. Scott, H. Ye, R. Henriquez, R. Crooks, Synthesis, Characterization, and Stability of Dendrimer-Encapsulated Palladium nanoparticles, *Chem. Mater.*, **15** (20), 3873-78 (2003).

[39] M. Peuckert, XPS Study on Surface and Bulk Palladium Oxide, its Thermal-stability, and a Composition with other Noble-Metal Oxides, *J. Phys. Chem.-US*, **89** (12), 2481-86 (1985).

[40] S. C. Tsang, P. J. F. Harris, M. L. H. Green, Thinning and Opening of Carbon Nanotubes by Oxidation Using Carbon-Dioxide. *Nature,* **362** (6420), 520-22 (1993).

[41] B. J. Hinds, N. Chopra, T. Rantell, R. Andrews, V. Gavalas, L. G. Bachas, Aligned multiwalled carbon nanotube membranes, Science, **303**, 62-65 (2004).

[42] P. M. Ajayan, T. W. Ebbesen, T. Ichihashi, S. Iijima, K. Tanigaki, H. Hiura, Opening Carbon Nanotubes with Oxygen and Implications for filling. *Nature,* **362** (6420), 522-25 (1993).

WELL ADHERED, NANOCRYSTALLINE, PHOTOACTIVE, TiO$_2$, THIN FILMS DIP-COATED ON CORONA-TREATED POLY(ETHYLENE TEREPHTHALATE) BY MODIFIED SOL-GEL PROCESSING AT ~95°C AND DRYING AT ~130°C

H.C. Pham[1], D.A.H. Hanaor[1], J.M. Cox[2], and C.C. Sorrell[1]

[1] School of Materials Science and Engineering, University of New South Wales
Sydney, NSW 2052, Australia
[2] School of Chemical Engineering, University of New South Wales
Sydney, NSW 2052, Australia

ABSTRACT

Nanocrystalline TiO$_2$ films were fabricated on 160 μm thick polyethylene terephthalate (PET) sheets and glass slides at temperatures below the boiling point of water. The modified sol-gel process used to produce the films involved (1) the preparation of a solution of titanium tetra-isopropoxide and propanol; (2) hydrolysis, peptization, and condensation in an aqueous solution of dilute nitric acid; and (3) refluxing at 95°C for 15 h to increase the crystallinity of the TiO$_2$. The resultant hybrid sol-gel suspension of nanoTiO$_2$ was deposited on corona-treated PET sheets by dip-coating at ~25°C, dried in air for 15-30 min, and dried in an oven at ~130°C for 15 min. It was dip-coated a second time, dried in air for 15-30 min, and dried in an oven at ~130°C for 15 h.

The characterization procedures were designed to provide information of the PET-coated samples in terms of the chemical composition and stoichiometry (XPS), film thickness and film/sheet interface (FIB), particle size and morphology (HRSEM and AFM), crystallinity and crystallite size (HRTEM on glass-coated samples), mineralogy (XRD and GAXRD), transparency (UV-VIS on glass-coated samples), photoactivity (methylene blue decomposition), and adhesion in quasi-tension and quasi-shear (tape tests).

The resultant films were (1) relatively thin (~570 nm), (2) of principally crystalline anatase with a small amount of brookite as secondary phase (both disordered), (3) of small anatase crystallite size (~4-7 nm) and larger rounded agglomerate size (~20-25 nm), (4) of relatively high transparency (~70-90%), (5) well adhered to the PET sheets (no damage from tape tests), and (6) photoactive (decomposing methylene blue).

The good adhesion was attributed to the establishment of covalent bonding between the TiO$_2$ film and PET substrate through processes similar to etherification and esterification involving the –OH ligands of the hydrolyzed ≡Ti–OH layer and the PET surface functionalized with ≡COH (etherification)[1] and/or –COOH (esterification)[2]. The hydrogen bonding between the two sets of –OH ligands initiated adhesion but the drying at 130°C caused dehydroxylation and the consequent establishment of CO–Ti covalent bonding.

INTRODUCTION

Titanium dioxide (TiO$_2$) has attracted considerable attention owing to its valence and conduction band levels and the associated ability to function as a semiconductor photocatalyst[3]. TiO$_2$ exhibits a photocatalytic effect through the photogeneration of electron hole pairs, known as *excitons*, upon exposure to radiation exceeding the material's band gap of ~3.2 eV[4,5]. These photo-generated holes and electrons can facilitate respectively the oxidative and reductive destruction of organic contaminants at the material's surface[6-8]. TiO$_2$ photocatalysis is particularly attractive for applications in water and air purification as these approaches enable the decomposition of pesticides and inactivation of bacteria and viruses in water streams without the consumption of reagents[9,10]. TiO$_2$ photocatalysts are used in different forms, including powder suspensions[11,12], nanostructured materials[13,14], thick films[15,16], and thin films[17-19]. The use of TiO$_2$ in thin-film form is particularly

31

attractive as the result of optical properties and the versatility in applications that can be used with photocatalysts; such applications include self-cleaning and self-sterilizing surfaces[20-23].

The crystalline structure of TiO_2 photocatalysts is of great importance in governing the performance of such materials. It is accepted that crystallinity is a requirement for photocatalytic activity in TiO_2[24,25] and the metastable anatase phase generally is considered to exhibit superior performance relative to the equilibrium rutile phase[26,27]. It often is reported that mixed-phase TiO_2 materials exhibit superior photoactivity through improved charge carrier separation[28-30].

The fabrication of TiO_2 photocatalysts in thin-film form can be achieved by a range of methods, including chemical vapour deposition (CVD)[8,31], physical vapour deposition (PVD)[20,32], and sol-gel coating (dip coating and spin coating)[33,34]. In most thin-film fabrication methods, high-temperature postdeposition treatment is required in order to achieve crystallisation of the films. It is for this reason that practical TiO_2 thin-film photocatalysts generally are limited to deposition on thermally stable substrates, such as metals and ceramics[4,35]. The development of low-temperature fabrication processes for TiO_2 thin films would enable their application on inexpensive and flexible polymeric substrates.

Low-temperature fabrication of crystalline TiO_2 thin films has been investigated using ion-assisted evaporation (at 150°C)[36], atomic layer deposition (at 165°-350°C)[37], and reactive pulsed sputtering (at 130°C)[38]. While these methods may generate thin films of high transparency and adhesive strength, the high cost of the necessary facilities and the requisite vacuum conditions limit their applicability. Partially crystalline TiO_2 thin films have been obtained using a number of solution-based procedures, including liquid-phase deposition[39] and layer-by-layer deposition[40]. However, such syntheses usually result in films of poor optical quality and adhesive strength.

The sol-gel process is a versatile technique for the fabrication of TiO_2 thin films owing to the relatively low cost and the potential to cover large substrates. These processes involve the controlled peptization and condensation (which must follow hydrolysis) of titanium-bearing precursors, typically alkoxides or inorganic titanium salts[25,41,42]. Conventional sol-gel thin-film fabrication procedures generally consist of hydrolysis, peptization, condensation, gelation, dehydration (drying), and high-temperature firing in order to obtain crystalline films. By this route, the temperatures required for crystallisation of anatase are in the range 450°-800°C[43,44]. These temperatures are well above the melting and decomposition temperatures of nearly all polymeric substrates. Film recrystallization also can occur during postdeposition autoclaving of as-prepared films at >100°C in high-pressure reactors[45].

Alternatively, the sol-gel procedure may be modified in order to obtain hybrid sol-gel suspensions containing TiO_2 nanocrystals before deposition, which therefore will result in crystalline TiO_2 films upon deposition. These hybrid suspensions of partially crystalline TiO_2 have been reported to be generated by promoting the complete hydrolysis of Ti-containing precursor and peptization in acidic media or by refluxing at low temperatures[46]. Such films, which demonstrate notable photocatalytic activity, have been deposited at low temperatures (~60°-140°C) on a range of substrates, including polymers[47-49]. Sánchez et al.[47] recently reported the modified sol-gel fabrication of crystalline TiO_2 films at 60°C on PET monoliths. In that work, crystallization of TiO_2 was promoted by acidic peptization and aging of the sol-gel hybrids at room temperature. Adhesion of the films to the polymer substrates was enhanced by addition of a fluorinated surfactant to the coating material and by precoating the polymer with poly(diallyl-dimethyl-ammonium chloride).

EXPERIMENTAL PROCEDURE

Sample Preparation

Melinex 516 polyester, a commercially available, biaxially-oriented, poly(ethylene terephthalate) (PET, DuPont Corp.), was used as the polymeric substrate. This product has a melting point of ~250°C and a highly crystalline structure, as confirmed by separate laser Raman microspectroscopy and X-ray diffraction (XRD) analyses. Prior to TiO$_2$ deposition, the PET sheets, in the form of ~25 mm x 80 mm x 160 μm films, were ultrasonically cleaned successively in liquid detergent solution, ethanol, acetone, and deionized water. They were placed in a drying oven and dried at 130°C for ~15 h. In order to activate the polymer surface and promote adhesion, the dried sheets were treated with a corona air plasma unit (BD-20AC, Electro-Technic) for ~1 min in air at 40 V, with a ~1 cm distance between electrode tip and PET surface. This resulted in partial oxidation; formation of oxidised functional groups, such as hydroxyl, carboxyl, *etc.*; and a resultant increase in surface energy[50,51].

The hybrid sol-gels were produced using a mixture of titanium tetra-isopropoxide (TTIP; 97 wt%, Sigma-Aldrich) and propanol (AlanaR, ≥99.7 wt%) added to an aqueous solution of nitric acid Univar, 70 wt%). The hydrolysis reaction of TTIP was moderated by mixing with the propanol.

The specific procedure consisted of the following steps. A volume of 25 mL of TTIP was added to 6 mL of propanol and then was mixed by manual stirring for ~1 min. In a second beaker, 150 mL of deionized water and 1.7 mL of concentrated nitric acid (Univar, 70 wt%) were combined and magnetically stirred for ~5 min. The solution of TTIP and propanol was transferred to a glass burette and then added dropwise over ~90 min to the aqueous solution of nitric acid with continuous magnetic stirring; this hydrolysis, peptization, and condensation procedure resulted in the formation of the hybrid sol-gel suspensions containing nanoTiO$_2$. A relatively dilute solution (H$_2$O/TTIP molar ratio ~83:1) was used in order to promote the complete hydrolysis of TTIP and peptization of the sols.

The sol-gel hybrid was transferred to a glass bottle with a loosely fitting lid, this was placed in a water bath at a controlled temperature of ~95°C, and the bath was placed on a magnetic stirring unit. The sol-gel hybrid was stirred for ~8 h in order to obtain complete peptization, facilitate Ostwald ripening[52], and remove propanol, at which point the volume had reduced to approximately half the original. At this point, sufficient deionized water was poured slowly into the bottle until the original volume was regained, after which the bottle was sealed with a tightly fitting ground glass stopper. The latter served as a pressure-relief system so that the subsequent ~8 h of reflux involved minimal (but allowable finite) loss of water vapour. No apparent volume decrease was visible after this second refluxing procedure. This process yielded translucent, hybrid, sol-gel suspensions that were stable for periods in excess of 6 months.

Thin films on PET were fabricated using two depositions with the aid of a motorised dip-coater using the same dipping and removal rate of 23 mm/min. The films were dried in air at ~25°C for ~15-30 min, after which they were placed in a drying oven and heated at 130°C for ~15 min. For the second coating, the dipping and drying processes were repeated using the same experimental conditions except that the final heating was done at 130°C for ~15 h. While most of the films were deposited on PET, those for high-resolution transmission electron microscopy (HRTEM, CM200-FEG, FEI, accelerating voltage 200 kV, holey carbon-coated copper grid, single-tilt specimen holder, liquid-nitrogen cooled) and ultraviolet-visible spectrophotometry (UV-VIS, Lambda 35, Perkin Elmer, single-beam, 200-700 nm) were deposited on more robust soda-lime-silica glass microscope slides. Further, for mineralogical analysis by X-ray diffraction (XRD, X'pert Materials Research Diffractometer, Philips, CuKα radiation, 45 kV, 40 mA, range 15°-70° 2θ, speed 0.03° 2θ/s, step size 0.01° 2θ), powders were produced by (1) drying at 25°C for 1 week, (2) refluxing at ~95°C and drying at 25°C for 1 week, (3) refluxing at ~95°C, (4) drying the hybrid sol-gel suspension in air at 25°C for ~24 h, and (5) heating at 130°C for ~15 h.

Sample Characterization

The surface chemistry and stoichiometry of each TiO$_2$ film was analysed by X-ray photoelectron spectroscopy (XPS, Thermo Scientific ESCALAB250Xi spectrometer, monochromated Al-K source [1486.6 eV], hemispherical analyser, multichannel detector).

The film thickness was determined and the appearance of the film/sheet interface was assessed using single-beam focused ion beam milling (FIB, XP200, FEI). The specimens initially were sputtered with a thin layer of gold (~20 nm) in order to enhance the electrical conductivity. A beam of gallium ions (Ga^{3+}) was used to mill holes of dimensions 20 μm x 6 μm x 2 μm. The cross-section of the film was analysed at an angle of 45°.

Microstructural features of larger scale, including cracks and pinholes, were examined using optical microscopy. The microscope was associated with a laser Raman microspectrometer (inVia, Renishaw, 50X).

The particle size and morphology of the films were assessed using high-resolution scanning electron microscopy (HRSEM, NovaNano SEM 230, FEI, accelerating voltage 2-3 kV, through-the-lens detector, spot size ~2 nm). In-lens secondary and backscattered electron modes were used.

The particle size and morphology were confirmed using atomic force microscopy (AFM, Nanoscope IIIA Multimode SPM, Veeco Instruments, tapping mode) under ambient conditions. The cantilever resonant frequencies were in the range 330-350 kHz and the force constant was ~42 N/m.

The crystallinity and crystallite size of the nanoTiO₂ were examined by HRTEM, as described.

The mineralogy of nanoTiO₂ was determined by XRD, as described. The mineralogy of the films was determined by glancing angle X-ray diffraction (GAXRD, X'pert Materials Research Diffractometer, Philips, 40 kV, 30 mA, angle of incidence 1° 2θ, range 20°-60° 2θ, speed 0.025° 2θ/min, step size 0.01° 2θ).

The optical absorption and transmission spectra of the films were determined by UV-VIS, as described.

The photoactivity was assessed by analysing the photodecomposition of an aqueous (deionized water) solution of methylene blue (MB, Sigma-Aldrich, 2 ppm by weight) under UV irradiation. The test samples consisted of TiO₂-coated PET films in the form of 25 mm x 30 mm x 160 μm sections that were TiO₂-coated on both sides. Each sheet was placed in a glass petri dish containing 10 mL MB solution. The dishes were covered with the lids, placed on an incubator rack, and uniformly illuminated from beneath using a weak UV illuminator (short-wave UVA at 365 nm) at an intensity of ~1 mW/cm² (determined by a Digitech QM1587 light meter). The distance between the UV light source and the sheet was ~3 cm. Blank samples in the form of uncoated PET films were subjected to the same procedure. After 5 h, the solution in each dish was collected, adjusted to 10 mL (typically with ~1 mL addition of deionized water), placed in a test tube, and sealed. The absorption spectra of the MB solutions were determined using UV-VIS, as described, except in the range 400-800 nm. The degradation of the MB was assessed by comparing the absorption spectra for solutions containing the coated sheets with those of the blanks.

The adhesion of the films was examined qualitatively using the adhesive tape test (Scotch Flatback Masking Tape 250, 3M). The coated PET sheet was secured to a glass microscope slide with a polyacrylate glue and the tape (250 mm x 300 mm area) was pressed tightly onto the sample by rubbing it by hand, with high downward force, with a pencil eraser for ~30 s. The quasi-tensile adhesion was assessed by peeling up ~5 mm of tape, gripping between thumb and forefinger, and rapidly pulling the tape from one end normal to the film surface. The quasi-shear adhesion was assessed by preparing the samples as described above and rapidly pulling the tape from one end parallel to the film surface.

RESULTS AND DISCUSSION

Film Chemistry

Figure 1a shows XPS spectra of the coatings, indicating that the binding energies of Ti2p3/2 and Ti2p3/1 were 458.9 eV and 464.7 eV, respectively. Since the binding energies were different from those of Ti metal (454.0 eV), TiO (455.0 eV), and Ti₂O₃ (456.7 eV), then the XPS data are consistent with those for TiO₂[53,54]. Moreover, the Ti2p doublet separation between the Ti2p3/2 and Ti2p3/1 peaks of ~5.7 eV, shown in Figure 1b, also is typical of stoichiometric TiO₂[54]. A clear peak for C and smaller peaks for Si (visible upon enlargement of the data plot) and N also were observed. The presence of a significant amount of C is ascribed to organic matter in the films from partially hydrolyzed TTIP solution and/or surface contamination[55]. The minute amounts of Si and N probably originated from the glass containers and residual nitric acid, respectively.

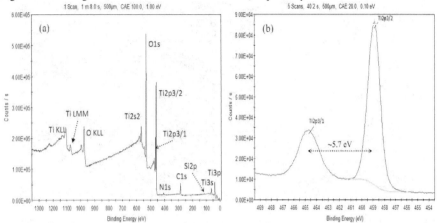

Figure 1. (a) XPS spectra for TiO$_2$ coating on PET and (b) enlargement of Ti2p doublet

Microstructural Analyses

To the eye, the TiO$_2$ coatings obtained on PET were transparent with weak spectral colour patterns, which resulted from interference fringes. This suggests that the film thickness was of the order of the wavelengths of visible light (380-740 nm), which was confirmed by FIB data that revealed film thicknesses of 570 ± 20 nm.

Under optical microscopy, as shown in Figure 2a, while no delamination was observed, the thin films contained well distributed pinholes and branched cracks in a limited number of areas, all of which were generated during drying. The increased surface area from these features can be expected to improve the photocatalytic activity but these features also are expected to decrease the adhesive strengths.

HRSEM images show that the films consisted of uniformly anhedral and equiaxed particles, with diameters in the range 15-25 nm, as shown in Figure 2b. A comparable study[48] using similar materials, processing procedures, and analytical methods yielded uniform TiO$_2$ particles of ~20 nm size. Since the HRSEM requires coating with a conducting layer of ~10 nm thickness (of, in this case, Cr), the indicated particle size range may be misleading. To clarify this issue, AFM analysis of uncoated films was done for comparison. These data, shown in Figure 3a, indicate that the Cr coating did not affect the data since the two particle sizes were approximately the same. It is apparent that these films have high surface areas, which are expected to enhance the photocatalytic activity through increased active surfaces sites and enhanced light capture through multiple reflection-absorption. These data also show that the individual TiO$_2$ grains are agglomerated and it is these agglomerates that are imaged in the HRSEM image.

Structural data for a single particle, obtained by HRTEM, are shown in Figure 3b. The observation of lattice fringes confirms that the grain shown is largely crystalline and that it is comprised of crystallites of ~4-7 nm in cross-section (outlined). Areas without lattice fringes are consistent with disordered (amorphous) regions between the crystallites. Consequently, the individual

TiO$_2$ grains comprising the agglomerates are themselves comprised of randomly oriented crystallites. The lattice fringe distance is ~0.35 nm, which may be contrasted with the lattice parameters of anatase of $a = 0.3785$ nm and $c = 0.9514$ nm[24].

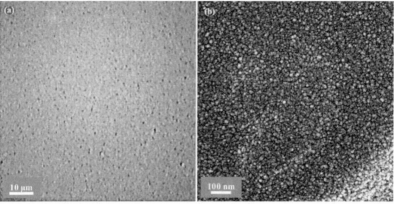

Figure 2. Micrographs of TiO$_2$ coating on PET: (a) by laser Raman (optical, at 50X) microscope and (b) by HRSEM

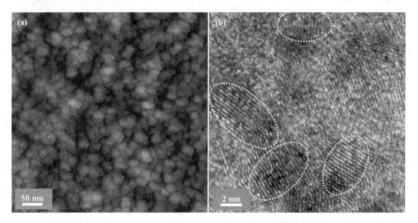

Figure 3. Micrographs of TiO$_2$ coating: (a) by AFM (on PET) and (b) by HRTEM (on glass)

Mineralogy

Figure 4 shows a typical GAXRD pattern of a TiO$_2$ film on PET. These data are consistent with the identification of the polymorph anatase. The broadening of the peaks relative to those observed in conventional X-ray powder diffraction usually is attributed to the quantum confinement effects caused by the small sizes of crystallites[56,57], a degree of crystallinity less than complete[58], and/or residual stress in the film[59]. The former two are most likely to be the cause of the nature of the peaks.

Figure 5 shows an XRD pattern of the dried powder, which was produced using a procedure essentially identical to that used to produce the films. These data confirm the identification of incompletely crystallized anatase, although they also reveal contamination by a small amount of poorly crystallized brookite, which has been observed by other researchers[48].

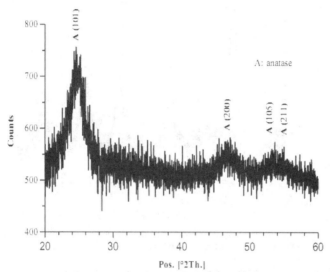

Figure 4. Glancing-angle X-ray diffraction pattern of TiO₂ coating on PET substrate

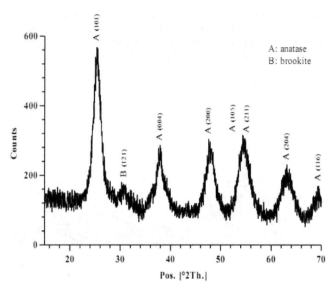

Figure 5. X-ray powder diffraction pattern of TiO$_2$ powder obtained refluxing at ~95°C for 15 h, drying in air at 25°C for ~24 h, and further drying at 130°C for ~15 h

The XRD data for the powder produced by refluxing at ~95°C and drying at 25°C for 1 week (not shown) were identical to those shown in Figure 5. This demonstrates that heating at 130°C was not responsible for the recrystallization of the anatase. Also, X-ray diffraction data for powder produced without refluxing at 95°C but only by drying for 1 week (also not shown) indicated that the anatase was only weakly crystalline. This demonstrates that the increase the crystallinity in order to generate an XRD pattern equivalent to that shown in Figure 5 is likely to depend on increased (1) hydrolysis, (2) peptization, and (3) Ostwald ripening[52] resulting from the refluxing at ~95°C for 15 h. Further, the inhibition of condensation until hydrolysis is well advanced or complete, by the use of nitric acid to achieve a pH of <1.5, generally is considered to be important. That is, if condensation precedes hydrolysis, undesirable amorphous material precipitates. However, when refluxing is used, the order of these two phenomena makes no difference – both yield reasonably well crystallized product. The necessity of such low-temperature heating by refluxing to ripen the particles has been reported by other researchers[48,52]. In contrast, the work of Sánchez et al.[47] shows that simultaneous hydrolysis, peptization, and condensation (which must be minimized in the absence of refluxing), followed by drying at only 60°C also can result in recrystallization of moderately well crystallized anatase.

Optical Properties

UV-VIS transmission spectra for a glass substrate and a glass substrate with deposited anatase film are shown in Figure 6 (the necessity of the use of a rigid substrate precluded the use of PET as substrate). A single-beam system was used in order to be able to accommodate the effects of differences in thickness and absorption between the substrate itself and the substrate with film (a dual-beam system subtracts the former data from the latter; it does not compare the data). These data reveal (1) a sharp absorption edge (~316 nm), (2) relatively high transmission (~70-90%), (3) a single partial interference fringe. While a sharp absorption edge is an advantage owing to the maximization of the range of absorbed wavelengths, the high transmission is a disadvantage owing to the converse low absorption. The interference fringe demonstrates that the film is very flat and of consistent thickness.

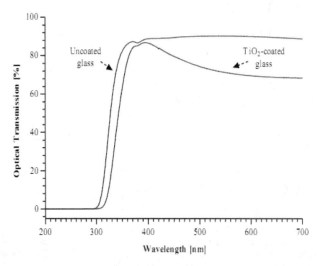

Figure 6. UV-VIS single-beam transmission spectra for glass substrate with and without TiO$_2$ coating

It may be noted that the small size of the crystallites may be responsible for a blue shift in the absorption edge to lower wavelengths. This effect would reduce the useful wavelengths that could be absorbed and used for photocatalysis. One study suggested that the effect becomes significant at crystallite sizes <2 nm[60] while another concluded that the equivalent effect was obtained with crystallite sizes <11 nm[61]. Since the crystallite sizes of 4-7 nm observed in the present work are between these sizes, then refluxing at higher temperatures and/or for longer times could increase the crystallite size to >11 nm and so preclude the potential for a blue shift.

Photocatalytic Properties

Figure 7 compares typical data for the UV-VIS absorption for aqueous methylene blue solutions in which were immersed uncoated PET control samples and PET coated with TiO$_2$ following irradiation by a weak UV light source for 5 h. These data confirm that the TiO$_2$ coatings exhibit clear photocatalytic activity.

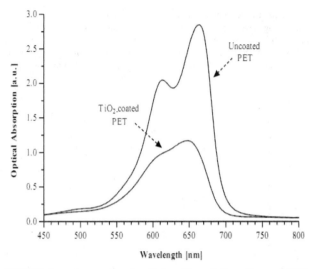

Figure 7. UV-VIS absorption spectra of methylene blue solution after weak UV irradiation

Film Adhesion

The present work demonstrates that thin films of crystalline TiO₂ can be deposited on corona-treated PET. The quasi-tensile and quasi-shear tape tests revealed that no damage to any of the tapes occurred by delamination, fracture, crumbling, or other mechanism. The good adhesion is likely to result not from the establishment of the expected hydrogen bonding between the ceramic and polymer phases but by the establishment of subsequent covalent bonding in a three-stage process:

Stage 1: The corona treatment removes the hydration layer initially present from atmospheric adsorption and replaces it with organic functional groups, such as =CO, ≡COH, and/or –COOH[50,51], as shown in Figure 8, on the surface of the PET. These groups are well bonded to the PET through carbon-carbon covalent bonding.

Stage 2: The TiO₂ film surface is hydrated and terminated by –OH groups, resulting in the presence of a ≡Ti–OH layer. The –OH ligands of the functional groups and of the ≡Ti–OH form hydrogen bonds and initiate adhesion.

Stage 3: Drying at 130°C causes dehydroxylation and consequent establishment of CO–Ti covalent bonding by processes similar to etherification (≡COH + OH–Ti≡) and/or esterification (–COOH + OH–Ti≡), thereby finalizing adhesion by covalent bonding. The process is summarised as follows:

Etherification: (PET)≡C–OH + HO–Ti≡(TiO₂ film) → (PET)≡CO–Ti≡(TiO₂ film) + H₂O

Esterification: (PET)–COOH + HO–Ti≡(TiO₂ film) → (PET)≡CO–Ti≡(TiO₂ film) + H₂O

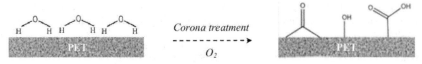

Figure 8. Effects of corona treatment in air on PET polymer surface

The good adhesion was attributed to the establishment of covalent bonding between the TiO$_2$ film and the PET substrate through processes similar to etherification and esterification involving the –OH ligands of the hydrolyzed layer and the PET surface functionalized with ≡COH (etherification) and/or –COOH (esterification). The hydrogen bonding between the two sets of –OH ligands initiated adhesion but the drying at 130°C caused dehydroxylation and the consequent establishment of CO–Ti covalent bonding.

CONCLUSIONS
Stable hybrid sol-gel suspensions of nanoTiO$_2$ can be synthesized at temperatures below the boiling point of water using a modified sol-gel route. The crystalline nanoTiO$_2$ was generated initially at room temperature during hydrolysis of the Ti-containing metallorganic precursor. The precipitation of additional particles and the crystallinity of the existing precipitates then were increased by refluxing for 15 h at 95°C. The crystallized TiO$_2$ consisted of crystallites of anatase and a small amount of brookite. The anatase was primarily in the form of grains comprised of crystallites sized ~4-7 nm, and these grains were agglomerated into secondary particles of size in the range ~20-25 nm.

Coatings of hybrid sol-gel suspensions onto corona-treated PET sheets can be fabricated by dip-coating, followed by drying at 130°C. The resultant transparent, nanocrystalline, anatase, thin films were well adhered to the substrates through processes similar to etherification and esterification, which resulted in the conversion of hydrogen bonding between –OH ligands, present on the TiO$_2$ and formed during corona treatment, to CO–Ti covalent bonding, formed during drying at 130°C.

The photoactivity of the nanoTiO$_2$ films was confirmed by the photodecomposition of aqueous solutions of methylene blue irradiated by UVA, which is the predominant form of UV radiation (90-95%) that reaches the earth's surface[62].

The present work shows that the fabrication of high-quality, photoactive, anatase, thin films can be achieved at temperatures well below those causing melting or pyrolysis of most polymers. This opens up the possibility of a simple and relatively low-cost means of production of technical devices, including flexible solar cells; photoanodes for photoelectrochemical decomposition of water; coatings for air and water purification, self-sterilization, and antifogging; and coatings for optical devices, particularly plastic lenses.

REFERENCES
[1]R. Barattin and N. Voyer, Chemical Modifications of AFM Tips for the Study of Molecular Recognition Events, *Chem. Commun.*, 1513-1532 (2008).
[2]Q. Junmin and L. Kathy, Multiwall Carbon Nanotube and TiO$_2$ Sol Assembly, *J. Nanosci. Nanotechnol.*, **9**, 5816-5822 (2009).
[3]N. Daude, C. Gout, and C. Jouanin, Electronic Band Structure of Titanium Dioxide, *Phys. Rev. B*, **15**, 3229-3235 (1977).
[4]G. Balasubramanian, D.D. Dionysiou, M.T Suidan, I. Baudin, and J.M. Lan, Evaluating the Activities of Immobilized TiO$_2$ Powder Films for the Photocatalytic Degradation of Organic Contaminants in Water, *Appl. Catal. B*, **47**, 73-84 (2004).

[5]K. Nagaveni, M.S. Hegde, N. Ravishankar, G.N. Subbanna, and G. Madrass, Synthesis and Structure of Nanocrystalline TiO_2 with Lower Band Gap Showing High Photocatalytic Activity, *Langmuir*, **20**, 2900-2907 (2004).

[6]J.M. Herrmann, Heterogeneous Photocatalysis: Fundamentals and Applications to the Removal of Various Types of Aqueous Pollutants, *Catal. Today*, **53**, 115-129 (1999).

[7]A. Linsebigler, G. Lu, and J.T. Yates, Photocatalysis on TiO_2 Surfaces: Principles, Mechanisms and Selected Results, *Chem. Rev.,* **95**, 735-758 (1995).

[8]A. Mills, N. Elliott, I.P. Parkin, S.A. O'Neill, and R.J. Clark, Novel TiO_2 CVD Films for Semiconductor Photocatalysis, *J. Photochem. Photobiol. Chem. A*, **151**, 171-179 (2002).

[9]A. Mills, R.H. Davies, and D. Worsley, Water Purification by Semiconductor Photocatalysis, *Chem. Soc. Rev.*, **22**, 417-434 (1993).

[10]R.L. Pozzo, M.A. Baltanas, and A.E. Cassano, Supported Titanium Oxide as Photocatalyst in Water Decontamination: State of the Art, *Catal. Today*, **39**, 219-231 (1997).

[11]Y.S. Chai, J.C. Lee, and B.W. Kim, Photocatalytic Disinfection of E. Coli in a Suspended TiO_2 Reactor, *Kor. J. Chem. Eng.*, **17**, 633-637 (2000).

[12]G.P. Fotou, S. Vemury, and S.E. Pratsinis, Synthesis and Evaluation of Titania Powders for Photodestruction of Phenol, *Chem. Eng. Sci.*, **49**, 4939-4948 (1994).

[13]R. Ma, K. Fukuda, T. Sasaki, M. Osada, and Y. Bando, Structural Features of Titanate Nanotubes/Nanobelts Revealed by Raman, X-Ray Absorption Fine Structure and Electron Diffraction Characterizations, *J. Phys. Chem. B*, **109**, 6210-6214 (2005).

[14]Y. Zhu, H. Li, Y. Koltypin, Y.R. Hacohen, and A. Gedanken, Sonochemical Synthesis of Titania Whiskers and Nanotubes, *Chem. Commun.*, 2616-2617 (2001).

[15]D. Hanaor, M. Michelazzi, J. Chenu, C. Leonelli, and C.C. Sorrell, The Effects of Firing Conditions on the Properties of Electrophoretically Deposited Titanium Dioxide Films on Graphite Substrates, *J. Eur. Ceram. Soc.,* **31**, 2877-2885 (2011).

[16]D. Hanaor, M. Michelazzi, P. Veronesi, C. Leonelli, M. Romagnoli, and C.C. Sorrell, Anodic Aqueous Electrophoretic Deposition of Titanium Dioxide using Carboxylic Acids as Dispersing Agents, *J. Eur. Ceram. Soc.*, **31**, 1041-1047 (2011).

[17]D. Hanaor, G. Triani, and C.C. Sorrell, Morphology and Photocatalytic Activity of Highly Oriented Mixed Phase Titanium Dioxide Thin Films, *Surf. Coat. Tech.*, **205**, 3658-3664 (2011).

[18]R.W. Matthews, Photooxidation of Organic Impurities in Water using Thin Films of Titanium Dioxide, *J. Phys. Chem.*, **91**, 3328-3333 (1987).

[19]U. Selvaraj, A. Prasadarao, S. Komarneni, and R. Roy, Sol–Gel Fabrication of Epitaxial and Oriented TiO_2 Thin Films, *J. Am. Ceram. Soc.*, **75**, 1167-1170 (1992).

[20]J.O. Carneiro, V. Teixeira, A. Portinha, A. Magalhaes, P. Countinho, and C.J. Tavares, Iron-Doped Photocatalytic TiO_2 Sputtered Coatings on Plastics for Self-Cleaning Applications, *Mat. Sci. Eng. B*, **138**, 144-150 (2007).

[21]A. Mills, S. Hodgen, and S.K. Lee, Self-Cleaning Titania Films: An Overview of Direct, Lateral and Remote Photo-Oxidation Processes, *Res. Chem. Intermed.*, **31**, 295-308 (2004).

[22]I.P. Parkin and R.G. Palgrave, Self-Cleaning Coatings, *J. Mater. Chem.*, **15**, 1689-1695 (2005).

[23]P. Hajkova, P. Spatenka, J. Horsky, I. Horska, and A. Kolouch, Photocatalytic Effect of TiO_2 Films on Viruses and Bacteria, *Plasma Process. Polym.*, **4**, 397-401 (2007).

[24]G. Tian, H. Fu, L. Jing, B. Xin, and K. Pan, Preparation and Characterization of Stable Biphase TiO_2 Photocatalyst with High Crystallinity, Large Surface Area, and Enhanced Photoactivity, *J. Phys. Chem. C*, **112**, 3083-3089 (2008).

[25]Q. Zhang, L. Gao, and J. Guo, Effects of Calcination on the Photocatalytic Properties of Nanosized TiO_2 Powders Prepared by $TiCl_4$ Hydrolysis, *Appl. Catal. B*, **26**, 207-215 (2000).

[26]D.A.H. Hanaor and C.C. Sorrell, Review of the Anatase to Rutile Phase Transformation, *J. Mater. Sci.*, **46**, 855-874 (2011).

[27]A. Sclafani and J. M. Herrmann, Comparison of the Photoelectronic and Photocatalytic Activities of Various Anatase and Rutile Forms of Titania, *J. Phys. Chem.*, **100**, 13655-13661 (1996).

[28]D.C. Hurum, A.G. Agrios, S.E. Crist, K.A. Gray, T. Rajh, and M.C. Thurnauer, Probing Reaction Mechanisms in Mixed Phase TiO$_2$ by EPR, *J. Electron Spectro. Relat. Phen.*, **150**, 155-163 (2006).

[29]D.C. Hurum, A.G. Agrios, K.A. Gray, T. Rajh, and M.C. Thurnauer, Explaining the Enhanced Photocatalytic Activity of Degussa P25 Mixed-Phase TiO$_2$ using EPR, *J. Phys. Chem. B*, **107**, 4545-4549 (2003).

[30]G. Li, L. Chen, M.E. Graham, and K.A. Gray, A Comparison of Mixed Phase Titania Photocatalysts Prepared by Physical and Chemical Methods: The Importance of the Solid-Solid Interface, *J. Mol. Catal. Chem. A*, **275**, 30-35 (2007).

[31]Z. Ding, X. Hu, G.Q.Lu, P.L. Yue, and P.F. Greenfield, Novel Silica Gel Supported TiO$_2$ Photocatalyst Synthesized by CVD Method, *Langmuir*, **16**, 6216-6222 (2000).

[32]D. Mardare, M. Tasca, M. Delibas, and G.L. Rusu, On the Structural Properties and Optical Transmittance of TiO$_2$ RF Sputtered Thin Films, *Appl. Surf. Sci.*, **156**, 200-206 (2000).

[33]E.D. Sam, M. Urgen, F.Z. Tepehan, and V. Gunay, Self-Cleaning Photoactive TiO$_2$ Coatings on SLS Glasses by Sol-Gel Dip-Coating, *Key Eng. Mat.*, **264**, 407-410 (2004).

[34]S.D. Sharma, D. Singh, K. Saini, C. Kant, V. Sharma, S.C. Jain, and C.P. Sharma, Sol-Gel Derived Super-Hydrophilic Nickel Doped TiO$_2$ Film as an Active Photocatalyst, *Appl. Catal. A*, **314**, 40-46 (2006).

[35]A. Fernandez, G. Lassaletta, V.M. Jimenez, A. Justo, A.R. Gonzalez-Elipe, J.M. Herrmann, H. Tahiri, and Y. Ait-Ichou, Preparation and Characterization of TiO$_2$ Photocatalysts Supported on Various Rigid Supports (Glass, Quartz and Stainless Steel). Comparative Studies of Photocatalytic Activity in Water Purification, *Appl. Catal. B*, **7**, 49-63 (1995).

[36]M. Sasase, S. Isobe, K. Miyake, T. Yamaki, and I. Takano, Surface Morphology of TiO$_x$ Films Prepared by an Ion-Beam-Assisted Reactive Deposition Method, *Thin Solid Films*, **281-282**, 431-435 (1996).

[37]J. Aarik, A. Aidla, T. Uustare, and V. Sammelselg, Morphology and Structure of TiO$_2$ Thin Films Grown by Atomic Layer Deposition, *J. Cryst. Growth*, **148**, 268-275 (1995).

[38]J. Sicha, D. Herman, J. Musil, Z. Stryhal, and J. Pavlik, High-Rate Low-Temperature DC Pulsed Magnetron Sputtering of Photocatalytic TiO$_2$ Films: The Effect of Repetition Frequency, *Nanoscale Res. Lett.*, **2**, 123-129 (2007).

[39]A. Dutschke, C. Diegelmann, and P. Lobmann, Preparation of TiO$_2$ Thin Films on Polystyrene by Liquid Phase Deposition, *J. Mater. Chem.*, **13**, 1058-1063 (2003).

[40]K.C. Krogman, N.S. Zacharia, D.M. Grillo, and P.T. Hammond, Photocatalytic Layer-by-Layer Coatings for Degradation of Acutely Toxic Agents, *Chem. Mater.*, **20**, 1924-1930 (2008).

[41]J. Chen, L. Gao, J. Huang, and D. Yan, Preparation of Nanosized Titania Powder via the Controlled Hydrolysis of Titanium Alkoxide, *J. Mater. Sci.*, **31**, 3497-3500 (1996).

[42]S. Mahshid, M. Askari, and M.S. Ghamsari, Synthesis of TiO$_2$ Nanoparticles by Hydrolysis and Peptization of Titanium Isopropoxide Solution, *J. Mater. Process. Tech.*, **189**, 296-300 (2007).

[43]J. Medina-Valtierra, M. Sánchez-Cárdenas, C. Frausto-Reyes, and S. Calixto, Formation of Smooth and Rough TiO$_2$ Thin Films on Fiberglass by Sol-Gel Method, *J. Mex. Chem. Soc.*, **50**, 8-13 (2006).

[44]X. Chen and S.S. Mao, Titanium Dioxide Nanomaterials: Synthesis, Properties, Modifications, and Applications, *Chem. Rev.*, **107**, 2891-2859 (2007).

[45]M. Langlet, A. Kim, and M. Audier, Liquid Phase Processing and Thin Film Deposition of Titania Nanocrystallites for Photocatalytic Applications on Thermally Sensitive Substrates, *J. Mater. Sci.,* **38**, 3945-3953 (2003).

[46]B.L. Bischoff and M.A. Anderson, Peptization Process in the Sol-Gel Preparation of Porous Anatase (TiO$_2$), *Chem. Mater.*, **7**, 1772-1778 (1995).

[47]B. Sánchez, J.M. Coronado, R. Candal, R. Portela, I. Tejedor, M.A. Anderson, D. Tompkins, and T. Lee, Preparation of TiO$_2$ Coatings on PET Monoliths for the Photocatalytic Elimination of Trichloroethylene in the Gas Phase, *Appl. Catal. B - Environ.*, **66**, 295-301 (2006).

[48]Y.J. Yun, J.S. Chung, S. Kim, S.H. Hahn, and E.J. Kim, Low-Temperature Coating of Sol–Gel Anatase Thin Films, *Mater. Lett.*, **58**, 3703-3706 (2004).

[49]M. Langlet, A. Kim, and M. Audier, Sol-Gel Preparation of Photocatalytic TiO$_2$ Films on Polymer Substrates, *J. Sol-Gel Sci.*, **25**, 223-234 (2002).

[50]M. Strobel, C.S. Lyons, J.M. Strobel, and R.S. Kapaun, Analysis of Air-Corona-Treated Polypropylene and Poly(Ethylene Terephthalate) Films by Contact-Angle Measurements and X-Ray Photoelectron Spectroscopy, *J. Adhes. Sci. Technol.*, **6**, 429-443 (1992).

[51]L.J. Gerenser, J.M. Pochan, J.F. Elman, and M.G. Mason, Effect of Corona Discharge Treatment of Poly(Ethylene-Terephthalate) on the Adsorption Characteristics of the Fluorosurfactant Zonyl FSC as Studied via ESCA and Surface-Energy Measurements, *Langmuir*, **2**, 765-770 (1986).

[52]G. Oskam, A. Nellore, R.L. Penn, and P.C. Searson, The Growth Kinetics of TiO$_2$ Nanoparticles from Titanium (IV) Alkoxide at High Water/Titanium Ratio, *J. Phys. Chem. B*, **107**, 1734-38 (2003).

[53]D. Huang, Z.D. Xiao, J.H. Gu, N.P. Huang, and C.W. Yuan, TiO$_2$ Thin Films Formation on Industrial Glass through Self-Assembly Processing, *Thin Solid Films*, **305**, 110-115 (1997).

[54]F. Zhang, S. Jin, Y. Mao, Z. Zheng, Y. Chen, and X. Liu, Surface Characterization of Titanium Oxide Films Synthesized by Ion Beam Enhanced Deposition, *Thin Solid Films*, **310**, 29-33 (1997).

[55]M.G. Liu, W. Jaegermann, J.J. He, V. Sundstrom, and L.C. Sun, XPS and UPS Characterization of the TiO$_2$/ZnPcGly Heterointerface: Alignment of Energy Levels, *J. Phys. Chem. B*, **106**, 5814-5819 (2002).

[56]A.I. Ekimov and A.A. Onushchenko, Quantum Size Effect in Three-Dimensional Microscopic Semiconductor Crystals, *JETP Lett.*, **34**, 345-349 (1981).

[57]P.F. Trwoga, A.J. Kenyon, and C.W. Pitt, Modeling the Contribution of Quantum Confinement to Luminescence from Silicon Nanoclusters, *J. Appl. Phys.*, **83**, 3789-94 (1998).

[58]O.M. Yaghi, H.L. Li, and T.L. Groy, Construction of Porous Solids from Hydrogen-Bonded Metal Complexes of 1,3,5-Benzenetricarboxylic Acid, *J. Am. Chem. Soc.*, **118**, 9096-9101 (1996).

[59]Z. Budrovic, H. Van Swygenhoven, P.M. Derlet, S. Van Petegem, and B. Schmitt, Plastic Deformation with Reversible Peak Broadening in Nanocrystalline Nickel. *Science*, **304**, 273-276 (2004).

[60]N. Satoh, T. Nakashima, K. Kamikura, and K. Yamamoto, Quantum Size Effect in TiO$_2$ Nanoparticles Prepared by Finely Controlled Metal Assembly on Dendrimer Templates, *Nature Nanotechnol.*, **3**, 106-111 (2008).

[61]A.J. Maira, J.M. Coronado, V. Augugliaro, K.L. Yeung, J.C. Conesa, and J. Soria, Fourier Transform Infrared Study of the Performance of Nanostructured TiO$_2$ Particles for the Photocatalytic Oxidation of Gaseous Toluene, *J. Catal.*, **202**, 413-420 (2001).

[61]S.Q. Wang, R. Setlow, M. Berwick, D. Polsky, A.A. Marghoob, A.W. Kopf, and R.S. Bart, Ultraviolet A and Melanoma: A Review, *J. Am. Acad. Dermatol.*, **44**, 837-846 (2001).

LARGE-SCALE SYNTHESIS OF MOS$_2$-POLYMER DERIVED SICN COMPOSITE NANOSHEETS

R. Bhandavat, L. David, U. Barrera, and G. Singh

Department of Mechanical and Nuclear Engineering, Kansas State University, Manhattan, KS 66506, USA

ABSTRACT

Bulk MoS$_2$ crystal is composed of vertically stacked and weakly interacting layers held together by van der Waals forces. Recently, MoS$_2$ nanosheets have been a topic of great interest particularly due to its tribological, semiconducting as well as lithium intercalation properties. Here, we demonstrate synthesis of single and few layer MoS$_2$ sheets using a liquid phase exfoliation process. The exfoliated flakes (few micrometers in lateral dimensions) were then interfaced with liquid polysilazane polymer to form MoS$_2$-polymer mix, which upon controlled heating formed MoS$_2$-SiCN composite. MoS$_2$-SiCN showed a sheet like morphology. Chemical structure and high temperature behavior of composite nanosheets was studied by use of various spectroscopic techniques and thermogravimetric analyzer, respectively. These composite nanosheets may find applications as a low coefficient of friction/ high temperature (~1200 °C) lubricant and/ or a durable negative electrode material for rechargeable batteries.

INTRODUCTION

Molybdenum disulfide or MoS$_2$ is a naturally occurring inorganic crystal composed of vertically stacked, weakly interacting layers held together by van der Waals interactions. In MoS$_2$, each Mo (IV) center occupies a trigonal prismatic coordination sphere, being bound to six sulfide ligands. Each sulfur center is pyramidal, being connected to three Mo centers. In this way, the trigonal prisms are interconnected to give a layered structure, where in molybdenum atoms are sandwiched between layers of sulfur atoms. Bulk MoS$_2$ is semiconducting with an indirect bandgap of 1.2 eV, whereas single-layer MoS$_2$ is a direct gap semiconductor with a band gap of 1.8 eV[1].

Several methods have been developed for synthesis of MoS$_2$; Du et al.[2] utilized a redox reaction to exfoliate MoS$_2$ that restacked turbostratically when dried and underwent an irreversible phase transition to disordered 2H-MoS$_2$ at temperatures above 98 °C. In 1994 Guay et al.[3] exfoliated MoS$_2$ by intercalating it with lithium and on reaction with water it formed H$_2$ gas that blew apart MoS$_2$ crystal into its constituent single molecular layers. More recently, Coleman et al.[4] utilized various organic solvents to produce a dispersion of commercially obtained MoS$_2$ powders. The resultant dispersions were centrifuged, and the supernatant was decanted. Most prominent results were obtained for N-vinyl-pyrrolidinone, N-methyl-pyrrolidinone (NMP), dimethyl-imidazolidinone and dimethyl acetamide (DMA).

Recent studies have also demonstrated synthesis of MoS$_2$-based hybrid nanocomposite materials for use in Li-ion battery anodes due to its planar structure and week van der Waals attraction that allow large amount of lithium intercalation[5]. But use of MoS$_2$ nanosheets in batteries and high temperature applications is limited due to (a) low success rate in separating individual or few layer flakes of MoS$_2$ in large quantities[1,4] and (b) its relatively low thermodynamic and chemical stability in air and moisture[6].

Recent advances in processing of polymer-derived ceramics (PDCs) have allowed their molecular level interfacing with carbon nanotubes to yield composite materials that are resistant to oxidation at high temperatures while still maintaining remarkable physical properties[7,8]. Similarly, since starting material for a PDC is a polymer (typically silazane or siloxane) in liquid form, it should be possible to disperse and functionalize MoS$_2$ nanosheets with the polymeric precursor and obtain MoS$_2$-ceramic composite nanosheets upon controlled heating.

Following this premise, we demonstrate exfoliation of MoS$_2$ into few layer MoS$_2$ in a variety of solvents and then functionalize it with polymer-derived silicon carbonitride (SiCN) through controlled thermolysis to yield MoS$_2$-SiCN composite that shows a sheet like morphology.

EXPERIMENT

Material preparation

MoS$_2$ dispersions were prepared in NMP (Fisher[TM]), iso-propanol (ISP) (Fisher[TM]) and DMA (Fisher[TM]). These dispersants were narrowed out after trying several commonly available laboratory solvents for composite synthesis. Briefly, 0.8 mg/mL of MoS$_2$ were exfoliated in solvent using bath sonication (Branson 2510) at 130 W for 1 hour. The top portion of the solution was taken out carefully and centrifuged at 1500 RPM for 90 minutes further allowing any remaining non-exfoliated sheets to settle down. This process is summarized in Figure 1(a) and (b).

For synthesis of MoS$_2$-PDC composite, 33 wt % of polyureamethylvinyl silazane (commercial name: Ceraset[TM]) was added slowly by stirring MoS$_2$ dispersion in NMP followed by drying at 80 °C in inert atmosphere. To form the ceramic composite, dried mix was first cross-linked at 400 °C for 2 hours and then pyrolyzed at 1000 °C for 4 hours in flowing nitrogen. These are typical processing parameters for PDCs requiring low temperature cross-linking of polymer precursors followed by high temperature thermolysis resulting in a ceramic that is chemically and thermodynamically stable up to ~1200 °C. A general reaction mechanism for the conversion of polyureamethylvinyl silazanes to SiCN ceramic is as follows:

Polyureamethylvinyl Silazane

Single phase amorphous
SiCN ceramic

(1)

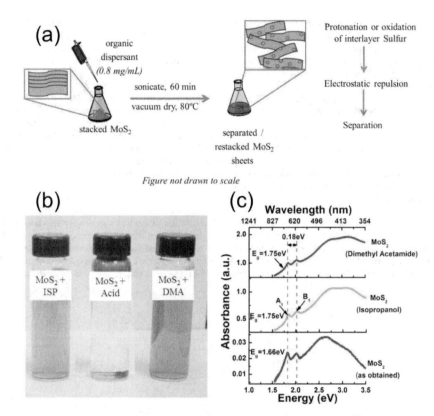

Figure 1. (a) Schematic of the MoS$_2$ exfoliation process with the insert showing separation of sheets and suggested exfoliation mechanism (b) Comparative suspension stabilities of MoS$_2$ dispersed in ISP, strong acid and DMA (c) Comparison of the optical absorbance spectra of "as-obtained" MoS$_2$ powder with dispersed MoS$_2$ nanosheets after centrifugation.

RESULTS AND DISCUSSION

Material Characterization

Scanning electron microscopy (SEM) was carried out using Carl Zeiss EVO 10 SEM and transmission electron microscopy (TEM) was performed on Philips CM 100 TEM (100 KeV). UV-vis absorption spectroscopy was performed on liquid specimens with Cary 500 Scan UV-vis NIR spectrometer using 1 cm quartz cuvettes from 800 nm to 200 nm using respective solvent as reference. Surface chemical composition of the powdered specimen was determined using X-ray photoelectron spectroscopy (XPS) which was performed using PHI Quantera SXM with Al Kα monochromatic X-radiation (beam size <9 μm) at 45° angle of incidence. Bruker X-ray diffractometer (XRD) with Cu-Kα radiation and nickel filter was used for the spectral collection of the finely crushed homogenous powdered composite specimen. Thermogravimetric analysis was performed by Shimadzu 50 TGA. Sample weighing, approximately 5 mg, was heated in a platinum pan at a rate of 10 °C/min in Helium gas flowing at 20 mL/min up to 1000 °C.

Uv-Vis Spectroscopy

Figure 1(c) compares the room temperature optical absorption spectra of dispersed MoS_2 nanosheets with the "as-obtained" MoS_2. The samples showed two sharp absorbance peaks at approximately 615 nm and 680 nm (2.01 eV and 1.82 eV) corresponding to the B1 and A1 excitonic transitions with the energy separation of ~ 0.18 eV. This is attributed to the spin-orbit splitting of the top of valence band at the K point in the Brillouin zone[9-11]. The third threshold at higher energy of 450 nm corresponds to a direct transition from valence band to the conduction band. Moreover, an increase of approximately 0.09 eV in the band gap for ISP and DMA dispersed MoS_2 was observed, suggesting exfoliation of nanosheets in these liquid solvents.

TEM and SEM

SEM images of the dispersed and dried specimen showed presence of large number of multi-layered MoS_2 sheets about 3 μm to 5 μm in size. Isolated cluster of stacked sheets were also observed with no major damage to the individual sheets. TEM images from the top portion of ISP dispersed specimen showed single layer or few layer MoS_2. Self-folding of individual sheets was also observed. Specimen prepared using NMP showed loosely stacked or restacked sheets probably due to post dispersion drying. Selected area electron diffraction (SAED) characterization showed presence of 2H-MoS_2.

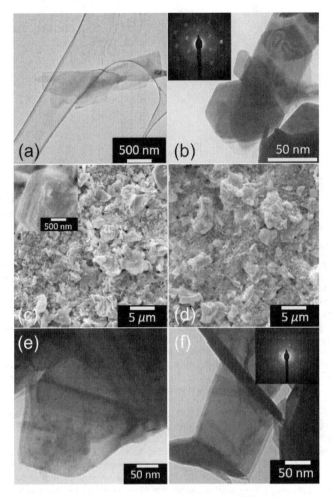

Figure 2. (a) and (b) TEM images of the MoS$_2$ sheets obtained from ISP and NMP dispersions, respectively. Insert is the SAED of hexagonal MoS$_2$ sheets. (c) and (d) SEM micrographs of MoS$_2$ and MoS$_2$-SiCN composite, respectively. (e) and (f) TEM images of MoS$_2$-SiCN composite sheets, respectively. Insert is the SAED showing crystalline MoS$_2$ in the composite.

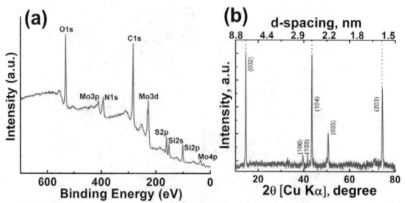

Figure 3. (a) X-ray photoelectron spectroscopy of the synthesized MoS₂-SiCN composite. (b) X-ray diffraction pattern of the pyrolyzed MoS₂-SiCN composite.

X-Ray Photoelectron Spectroscopy

The surface chemical characterization of synthesized MoS₂-SiCN composite showed presence of characteristic Mo, S, Si, C and N peaks. Deconvolution of individual elemental peaks indicated presence of Mo-Si bonds suggesting epitaxial functionalizing of MoS₂ sheets by SiCN ceramic domains and Mo-O peaks implying the reacted/ functionalized sidewalls of the nanosheets.

X-Ray Diffraction

XRD spectra of the crushed MoS₂-SiCN crushed powder sample indexed using Bragg's law revealed intense peaks at 2θ degrees of 14.5, 43.4 and 74.3 signifying (002), (104) and (203) lattice planes [JCPDS card no. 6-0097] characteristic of hexagonal MoS₂. The highest intensity (prominence) of these planes probably signifies the lattice plane's preferred orientation in the synthesized nanocomposite. No Si/C/N peaks were observed, this was not surprising considering that polymer-derived ceramics are predominantly amorphous when processed under ~1200 °C[12].

Thermogravimetric Analysis

Thermodynamic stability of the synthesized composite at high temperature was accessed by carrying out TGA at a heating rate of 10 °C/min. Similar to previous studies, the "as-obtained" MoS₂ specimen exhibited maximum weight loss in the 750 °C to 800 °C range leaving negligible (~6 %) residue. Whereas, the SiCN functionalized MoS₂ nanosheets showed excellent stability in the RT to 600 °C temperature range. The maximum weight loss was observed at 820 °C with approximately 34 % residue. The weight loss for the composite specimen is believed to be due to non-functionalized MoS₂ since PDCs are stable up to 1000 °C. Assuming a typical polymer to ceramic yield of 60 % and with the known behavior of "as-obtained" MoS₂ TGA we estimate the amount of MoS₂ in residual composite to be approximately 15 %. This suggests partial protection of MoS₂ in SiCN surroundings at high temperatures.

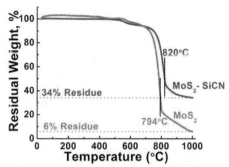

Figure 4. TGA comparison of the high temperature thermal stability of the as obtained MoS₂ with the MoS₂-SiCN composite

CONCLUSION

We demonstrated exfoliation of bulk MoS₂ into few layers MoS₂ sheets using a variety of organic solvents of which ISP showed the best results. The exfoliated nanosheets were then functionalized with PDC-SiCN through controlled thermolysis of a liquid polysilazane precursor. Chemical structure and optical behavior of the flakes was studied by use of XRD, XPS and UV-vis. Successful functionalization of MoS₂ sheets with the ceramic was confirmed by presence of Mo-Si type bonds in XPS. Preference of planar orientation of MoS₂ sheets in the composite could be interpreted from the XRD spectra. TGA indicated increased thermal stability of the MoS₂ sheets due to SiCN functionalization. Future work will address use of composite material as a low coefficient of friction/ high temperature (~1200 °C) lubricant and a durable negative electrode material for rechargeable batteries.

ACKNOWLEDGEMENTS

This research is based upon work supported by the National Science Foundation under grant no. EPS-0903806 and the State of Kansas through Kansas Technology Enterprise Corporation. Authors would like to thank Dr. Jerry Hunter for help with XPS and Professor Kenneth Klabunde at K-State for TGA usage.

REFERENCES

[1]B. Radisavljevic, A. Radenovic, J. Brivio, V. Giacometti and A. Kis, Single-layer MoS(2) transistors, *Nature Nanotechnology*, 6, 147-50 (2011).
[2]G. Du, Z. Guo, S. Wang, R. Zeng, Z. Chen and H. Liu, Superior stability and high capacity of restacked molybdenum disulfide as anode material for lithium ion batteries. *Chemical Communications*, **46,** 1106-8 (2010).
[3]D. Guay, W.M.R. Divigalpitiya, D. Belanger and X.H. Feng, Chemical Bonding in Restacked Single-Layer Mos(2) by X-Ray-Absorption Spectroscopy, *Chemistry of Materials*, **6,** 614-9 (1994).
[4]J.N. Coleman, M. Lotya, A. O'Neill, et al, Two-Dimensional Nanosheets Produced by Liquid Exfoliation of Layered Materials, *Science*, **331,** 568-71 (2011).

[5]K. Chang and W. Chen, L-Cysteine-Assisted Synthesis of Layered MoS(2)/Graphene Composites with Excellent Electrochemical Performances for Lithium Ion Batteries, *ACS Nano*, **5,** 4720-8 (2011).
[6]J.K. Lancaster, A Review of the Influence of Environmental Humidity and Water on Friction, Lubrication and Wear, *Tribol. Int.*, **23,** 371-89 (1990).
[7]J.H. Lehman, K.E. Hurst, G. Singh, E. Mansfield, J.D. Perkins and C.L. Cromer, Core-shell composite of SiCN and multiwalled carbon nanotubes from toluene dispersion, *J. Mater. Sci.*, **45,** 4251-4 (2010).
[8]L.N. An, W.X. Xu, S. Rajagopalan, et al, Carbon-nanotube-reinforeed polymer-derived ceramic composites, *Adv Mater*, **16,** 2036,+ (2004).
[9]J.P. Wilcoxon and G.A. Samara, Strong Quantum-Size Effects in a Layered Semiconductor - Mos2 Nanoclusters, *Physical Review B*, **51,** 7299-302 (1995).
[10]G. Eda, H. Yamaguchi, D. Voiry, T. Fujita, M. Chen and M. Chhowalla, Photoluminescence from Chemically Exfoliated MoS(2), *Nano Letters*, **11,** 5111-6 (2011).
[11]J. Brivio, D.T.L. Alexander and A. Kis, Ripples and Layers in Ultrathin MoS2 Membranes, *Nano Letters*, **11,** 5148-53 (2011).
[12]P. Colombo, G. Mera, R. Riedel and G.D. Soraru, Polymer-Derived Ceramics: 40 Years of Research and Innovation in Advanced Ceramics, *J Am Ceram Soc*, **93,** 1805-37 (2010).

SYNTHESIS OF TiO$_2$/SnO$_2$ BIFUNCTIONAL NANOCOMPOSITES

Huaming Yang [*] and Chengli Huo
Department of Inorganic Materials, School of Resources Processing and Bioengineering,
Central South University
Changsha 410083, China

* Corresponding author. hmyang@csu.edu.cn (H. Yang).

ABSTRACT

TiO$_2$/SnO$_2$ nanocomposites with different Ti/Sn molar ratios have been successfully prepared via the sol-gel route. The crystal size of the as-synthesized TiO$_2$/SnO$_2$ nanoparticles is about 80 nm. XRD analysis showed that the diffraction peaks associated with SnO$_2$ can't be found in XRD patterns when the molar ratio of TiO$_2$/SnO$_2$ was less than 10/1. The photocatalytic degradation of methyl orange (MeO) in TiO$_2$/SnO$_2$ suspension was investigated. The results indicated that the TiO$_2$/SnO$_2$ nanopcomposites had higher photocatalytic activity than pure TiO$_2$ when the molar ratio of TiO$_2$/SnO$_2$ was 10/1, but it will remarkably decrease with TiO$_2$/SnO$_2$ molar ratio of 1/1. In addition, the conductive property of TiO$_2$/SnO$_2$ nanocomposites suggested an inhibition electric effect at higher TiO$_2$ content. TiO$_2$/SnO$_2$ nanocomposites with Ti/Sn molar ratio of 1/10 showed a resistivity of 1.4×10^3 Ω·cm, which indicated a potential application in photocatalytic and electronic materials based on its bifunctional characteristics.

INTRODUCTION

In recent years, numerous efforts have been attempts to improve the photocatalytic activity of TiO$_2$ photocatalysts by modifications, such as doping, codeposition of metals, surface chelation, mixing of two semiconductors, etc.[1,2] Since the presence of a second semiconductor can improve the charge separation in the photocatalytic process.[3] Particularly, the TiO$_2$/SnO$_2$ system has shown a good performance for the degradation of pollutants.[4]

Tin oxide (SnO$_2$) is a wide band gap (E$_g$=3.6eV) n-type semiconductor used in many applications such as gas sensors and as anode material in Li-based batteries. Usually, this material is crystallized in rutile (TiO$_2$) structure. Theoretically, it can be excited by photons with the wavelengths below 326nm, but it shows only low photocatalytic activity under UV light in the present experimental conditions. The resistivity and dielectric properties are two important factors to characterize SnO$_2$ semiconductor.[5] The addition of TiO$_2$ can improve the electric properties of SnO$_2$. In this paper, we prepared SnO$_2$/TiO$_2$ nanocomposites with different Sn/Ti ratio, which have diplex properties, photocatalytic activity and electric property.

EXPERIMENTAL

The starting materials were AR-grade Ti(OBu)$_4$, SnCl$_2$·6H$_2$O, ammonia and anhydrous alcohol. SnCl$_2$·6H$_2$O and Ti(OBu)$_4$ dissolved in the anhydrous alcohol mixed as different ratio according to the requirement of SnO$_2$/TiO$_2$ nanocomposites, ultrasonically dispersed to form a mixture. And then a little amount of deionized water was slowly dripped into the mixture, which was stirred at the same time. pH value of solution was kept to be 10.0, controlled by NH$_3$·H$_2$O. The solution was aged at ambient temperature, followed by filtering, washing for several times with deionized water and anhydrous alcohol and drying at 100°C for 12h. And a precursor was obtained. Subsequent calcination of the precursor at 500°C for 3h in air results in the formation of nanocrystalline TiO_2 and SnO_2.

RESULTS AND DISCUSSION

Figure 1 shows the X-ray patterns of (a) TiO_2, (b) $TiO_2:SnO_2=10:1$, (c) $TiO_2:SnO_2=1:1$, (d) $TiO_2:SnO_2=1:10$ and (e) SnO_2 calcined at 500°C for 3h. As can be seen, the phase composition varies with the increasing content of SnO_2 and the intensity of the peaks associated with SnO_2 increased to some extent. At the same time, the addition of SnO_2 can trigger the anatase-rutile transformation, as reported previously.[6,7] As shown in Figure 1, when the amount of SnO_2 doped is less than 9 wt%, the diffraction peaks of SnO_2 cannot be found in XRD patterns. Chen et al. thought that this illustrates that SnO_2 is highly dispersed in the bulk phase of the nanocomposites.[8] When the amount of SnO_2 doped become high, the diffraction peaks of SnO_2 can be found in XRD patterns. Since no new crystal phases are found, it can be concluded that a new solid is not formed in the process of preparing SnO_2/TiO_2 nanocomposites. It is known from the calculation of the Scherrer's equation that the diameter of the composite oxides is not obviously changed. The average crystal size is about 80 nm.

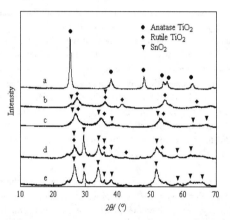

Figure 1. XRD patterns of SnO_2/TiO_2 nanocomposites
(a-TiO_2; b-$TiO_2:SnO_2=10:1$; c-$TiO_2:SnO_2=1:1$; d-$TiO_2:SnO_2=1:10$; e-SnO_2)

The photocatalytic oxidation of methyl orange in SnO_2/TiO_2 suspension under UV illumination was investigated in order to evaluate the photocatalytic activity of the as-prepared SnO_2/TiO_2 nanoparticles. A 150 ml of a 3.2×10^{-3} mol/l MeO solution, which characteristic absorption wavelength was 471 nm, was employed as a target. The mixture inside a 500 ml beaker remained in suspension by magnetic stirring. A 125 W high-pressure mercury lamp (GYZ-125) fixed at a distance of 16cm above the surface solution was used as a UV light source. And the pH value was kept to be 1.0. The of the MeO solution, which was drawn and then centrifuged to remove the semiconductor every was measured with a UV-vis spectrophotometer. The degree of MeO decolorization could be calculated according to the equation $C = (A_0 - A)/A_0 \times 100\%$, where C is the decolorization degree, A_0 is the initial absorbance of methyl orange solution, and A is the absorbance of the methyl orange solution after photocatalysis. In fact, there exists a linear relationship between the absorbance and concentration of the MeO solution under the same condition in our experiment. Therefore, the degree of MeO decolorization indicated its photodegradation.

Figure 2 shows the relationship between the degradation degree of MeO and the SnO₂/TiO₂ nanocoposites with different SnO₂-doped amounts. As shown in Figure 2, when the molar ratio of TiO₂:SnO₂ is 10:1, the photocatalytic activity of SnO₂/TiO₂ nanocoposites is higher than pure TiO₂, but it will decrease remarkably with doped- SnO₂ equal to TiO₂. In Ref [9], SnO₂/TiO₂ photocatalysts also show that the activity of the photocatalyst increases with increasing the amount of doped SnO₂.

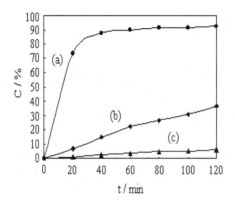

Figure 2. Effect of different SnO₂-doped amounts on the photocatalytic decolorization of the MeO solution (a-TiO₂/SnO₂=10/1; b-TiO₂; c-TiO₂/SnO₂=1:1)

When SnO₂ and TiO₂ form a coupled photocatalyst, as the electric potential of SnO₂ conduction band is about E_{CB}=0.45eV and TiO₂ about –0.5eV (the potential is relative to hydrogen electrode, pH 1), the CB of SnO₂ is lower than that of TiO₂, the former can act as a sink for the photogenerated electrons [8]. When TiO₂ and SnO₂ are excited simultaneously under UV illumination, the photogenerated electrons of the TiO₂ conduction band will be transferred to the conduction band of SnO₂. Since the holes move in the opposite direction from the electrons and at the same time photogenerated holes might be trapped within the TiO₂ particle, which make charge separation more efficient, resulting in the SnO₂/TiO₂ coupled photocatalyst exhibiting even higher photocatalytic activity than that of TiO₂. When TiO₂ is only excited by UV, the electron-hole pairs are produced on the TiO₂ surface, photogenerated electrons will be transferred toward SnO₂ conduction band, while holes will remain in the valence-band of TiO₂ and then be captured to take part in the reaction, which make charge separation more efficient, and thus the SnO₂/TiO₂ photocatalyst shows higher photocatalytic activity. But when the doped-SnO₂ amount is excessive, the properties will invalidate obviously.

Table 1 shows the SnO₂/TiO₂ nanocomposites with different contents of TiO₂. The resistivity is minimum when the amount of doped TiO₂ is 9 wt%. At 50% TiO₂, the resistivity of SnO₂/TiO₂ nanocomposites is maximum, suggesting an inhibition electric effect at higher TiO$_2$ content. SnO$_2$ doped with TiO$_2$ can form a transpositional solid solution, which has different degree of structural defects. Due to the increase of TiO$_2$-doped amounts, oxygen hole concentration on the surface of SnO$_2$/TiO$_2$ nanocomposites increases. Electron transfer becomes easier and the conductivity is improved correspondingly. However, when the doped-TiO$_2$ amount is excessive, ionic radius of Sn^{4+} and Ti^{4+} are slightly equivalent, oxygen hole concentration decreases and the conductivity of SnO$_2$/TiO$_2$ nanocomposites is receded.

Table I   The resistivity of SnO$_2$/TiO$_2$ nanocomposites

Amount of doped TiO$_2$ / %	0	9	50
Resistivity ρ / ($\times 10^3 \, \Omega \cdot$cm)	3.55	1.40	4.22

CONCLUSION

In summary, SnO$_2$/TiO$_2$ nanocomposites as prepared in our experiment have diplex properties, photocatalytic activity and electric property. The doped of SnO$_2$ into TiO$_2$ improve the photocatalytic activity of pure TiO$_2$ and the doped of TiO$_2$ into SnO$_2$ can improve the conductivity of pure SnO$_2$. An important point is that the excessive amount of the adulterants will inhibit the effects.

ACKNOWLEDGEMENTS

This work was supported by the 863 Program (2007AA06Z121) and the National Natural Science Foundation of China (50774095).

REFERENCES

[1] J. Bandara, C.C. Hadapangoda, and W.G. Jayasekera, TiO$_2$/MgO Composite Photocatalyst: the Role of MgO in Photoinduced Charge Carrier Separation, *Appl. Catal. B*, **50**, 83-88 (2004)

[2] S. Sato, R. Nakamura, and S. Abe, Visible-light Sensitization of TiO$_2$ Photocatalysts by N Doping, *Appl. Catal. A*, **284**, 131–137. 2005,

[3] F. Fresno, C. Guillard, and J.M. Coronado, Photocatalytic Degradation of a Sulfonylurea Herbicide Over Pure and Tin-Doped TiO$_2$ Photocatalysts, *J. Photoch. Photobio. A*, **173**, 13-20 (2005)

[4] G.M. Zuo, Z.X. Cheng, and H. Chen, Study on Photocatalytic Degradation of Several Volatile Organic Compounds, *J. Hazard. Mater.*, **128**, 158-163 (2006)

[5] P. Thangadurai, A. Chandra Bose, and S. Ramasamy, High Pressure Effects on Electrical Resistivity and Dielectric Properties of Nanocrystalline SnO$_2$, *J. Phys. Chem. Solids*, **66**, 1621-1627 (2005)

[6] F. Fresno, J.M. Coronado, and D. Tudela, Influence of the Structural Characteristics of Ti$_{1-x}$Sn$_x$O$_2$ Nanoparticles on Their Photocatalytic Activity for the Elimination of Methylcyclohexane Vapors, *Appl. Catal. B.*, **55**, 159-167 (2005)

[7] G.B. Song, J.K. Liang, and F.S. Liu, Preparation and Phase Transformation of Anatase-Rutile Crystals in Metal Doped TiO$_2$/Muscovite Nanocomposites, *Thin Solid Films*, **491**, 110-116 (2005)

[8] S. Chen, L. Chen, and S. Gao, The Preparation of Coupled SnO$_2$/TiO$_2$ Photocatalyst by Ball Milling, *Mater. Chem. Phys.*, **98**, 116-120 (2006)

[9] N. Kanai, T. Nuida, and K. Ueta, Photocatalytic Efficiency of TiO$_2$/SnO$_2$ Thin Film Stacks Prepared by DC Magnetron Sputtering, *Vacuum*, **74**, 723-727 (2004)

FABRICATION OF POROUS MULLITE BY FREEZE CASTING AND SINTERING OF ALUMINA-SILICA NANOPARTICLES

Wenle Li, Margaret Anderson, Kathy Lu
Materials Science and Engineering Department

John Y. Walz
Chemical Engineering Department
Virginia Polytechnic Institute and State University
Blacksburg, Virginia, USA

ABSTRACT

Porous mullite with interconnected porous microstructure is fabricated by freeze-casting and sintering of alumina-silica nanoparticles. The initial alumina and silica nanoparticle concentrations that are the same as mullite composition ($3Al_2O_3 \cdot 2SiO_2$) lead to the finest microstructure and the highest porosity (~69%) compared to either alumina-rich or silica-rich starting composition. Upon sintering, alumina and silica nanoparticles interact with each other and diffuse together to form the mullite phase at temperatures 1300°C or higher. The flexural strength of the composite keeps increasing with the increase of the sintering temperature. The fabricated porous mullite possesses improved strength than the porous alumina-silica composite.

INTRODUCTION

Porous ceramics are used in a wide variety of applications such as catalyst carriers, liquid and gas filters, sensors, and artificial bones because of their adjustable porous microstructure, high specific surface area, and thermal/chemical/mechanical stabilities.[1,2,3,4] There are many methods to produce porous structures, including gel casting, combustion processing, reaction forming, and spray drying.[4,5,6,7] Gel casting utilizes the interconnected liquid networks to create porous materials once the liquid is removed by the subsequent drying process.[8,9] However, the formation of cracks associated with the volume shrinkage during the drying process is a general issue in this technique.[10] In addition, this technique is limited to particular systems that a gel formation process is required. Freeze-casting technique has recently drawn much attention as it overcomes the above limitations and is a versatile method that can be applied to most colloidal suspensions. The typical freeze-casting technique includes three steps: preparation of well-dispersed suspension, solidification of the dispersing medium, and solidified medium sublimation.[11] The freeze-cast samples can be further sintered to the desired density and microstructure. This approach is desirable for making materials with adjustable porous structures and complex shapes. The designed porous microstructure can be fabricated by carefully controlling freezing temperature, freezing rate, solids loading, temperature gradient, additives, and applying external fields.

Mullite (also known as porcelainite, $3Al_2O_3 \cdot 2SiO_2$) is a silicate mineral which belongs to post-clay genesis. There are two typical particle shapes of mullite, plate-like and needle-like. The latter is of particular significance in improving the mechanical properties and the thermal shock resistance of the ceramics. So far, a lot of work has been devoted to preparing mullite, such as combustion of aluminum nitrate and silica fume,[5] sintering of Al_2O_3 and Si_3N_4,[3] and directly fusing Al_2O_3 and SiO_2 particles.[4] Mullite, whose melting temperature is 1840°C, is found to typically form at temperatures above 1400°C.[2,12,13] With the advancement of nanoparticle material sintering, the formation temperature can be further lowered as the nano-sized particles enhance the diffusion and activity of the sintering system.[14,15,16] It is also well known that the loss of strength associated with the porosity increase is a general issue for porous materials.[17] Thus, the fabrication of porous mullite by Al_2O_3 and SiO_2 nanoparticles is a promising technique to lower the mullite formation temperature and increase the strength of the resultant porous material.

The aim of the current study is to fabricate porous mullite by freeze-casting and sintering Al_2O_3 and SiO_2 nanoparticles. The phase development as a function of sintering temperature is studied to identify the mullite formation. Initial compositions with different Al_2O_3-to-SiO_2 ratios are investigated to determine the best composition for the porous structure. The strengthening effect of the developed mullite phase on the porous material is analyzed by a flexural strength test.

EXPERIMENTAL PROCEDURE

Raw Materials and Sample Preparation

In this study, a well-dispersed colloidal suspension was first prepared by mixing 34 wt% silica nanoparticle suspension (Ludox TMA, Sigma Aldrich, St. Louis, MI), de-ionized water, alumina nanoparticles (Alfa Aesar, Ward Hill, MA), sodium chloride (AR grade, Mallinckrodt, Paris, KY), and sodium hydroxide (Mallinckrodt, Paris, KY). The nominal diameter and density for silica nanoparticles were 22 nm and 2.37 g/cm^{3},[18,19] respectively. The alumina nanoparticles possessed a nominal diameter of 37.5 nm and a density of 3.6 g/cm^{3}. Sodium hydroxide was used to adjust the suspension pH to facilitate the gel formation while sodium chloride was introduced to generate the gelation. After the suspension was homogeneously mixed, it was poured into round silicone molds with a diameter of 31 mm and a depth of 2 mm. The resulting sample volume is approximately 0.15 cm^3. The samples were rest for 30 min to allow gel formation. Then the gels (within the containers) were placed directly into an Advantage Freeze Dryer (SP Industries VirTis, Gardiner, NY) for freeze casting. The samples were frozen and dried at -35°C for 700 min with a cooling rate of 2°C/min.

After freeze-casting, the fabricated green samples were sintered at 1200°C and 1300°C for 1 h with fixed heating and cooling rates of 5°C/min. The holding time was designed to be 1 h with the purpose of maintaining a high porosity while improving the mechanical strength.

In this study, three initial compositions (i.e. alumina-to-silica molar ratio: 3-to-3, 3-to-2, and 3.4-to-2) were used at a fixed solids loading of 18 vol%. This solids loading was chosen in consideration of the gel stability. Since the total solids loading was fixed, the increase of alumina concentration meant the decrease of silica concentration. The 3-to-2 alumina-to-silica ratio is the stoichiometry of mullite. Therefore, 3-to-3 and 3.4-to-2 compositions corresponded to silica-rich and alumina-rich compositions, respectively. Again, these stoichiometries were chosen to maintain the gel stability. At 18 vol% solids loading, the alumina volume fractions for 3-to-3, 3-to-2, and 3.4-to-2 alumina-to-silica ratio compositions were 9.51 vol%, 11.28 vol%, and 11.80 vol%. From now on, the different compositions will be noted based on the alumina concentration.

Characterization

The microstructures of the fabricated materials were observed using a field emission scanning electron microscope (SEM, LEO1550, Carl Zeiss MicroImaging Inc., Thornwood, NY). The samples were coated with a 3-5 nm Au-Pt layer before examination. Images were taken on the interior (cross-section) surfaces of the composite samples (the cross-sections were obtained by breaking the samples by hand and imaging one of the exposed surfaces). The porosity of the sample (void volume/sample volume) was calculated using the known volumes of components used to create the samples and the actual sample volume measured using a caliper and actual weight measured using a balance. The X-ray diffraction pattern was taken to study the phase development during the sintering process by an X'Pert PRO diffractometer (PANalytical B.V., EA Almelo, The Netherlands). The equibiaxial flexural strength of the freeze-cast composites was measured by a strength test apparatus with a 1 kN load cell (Instron 4204, Instron, Norwood, MA). The test was a two-dimension bending compressive strength measurement with 360° rotation about the load axis, which followed ASTM C 1499.[20] The crosshead was lowered at a speed of 0.1 mm/min until the sample was broken. The equibiaxial flexural strength was calculated as

$$\sigma_f = \frac{3F}{2\pi h^2}\left[(1-v)\frac{D_S^2-D_L^2}{2D^2}+(1+v)\ln\frac{D_S}{D_L}\right] \qquad (1)$$

where σ_f is the equibiaxial flexural strength, F is breaking load, h is specimen thickness, v is Poisson's ratio, D is sample diameter, D_S is support ring diameter, and D_L is load ring diameter. The Poisson's ratios for porous materials, v_p, were obtained using the following equation of Arnold et al.[21]

$$v_p = 0.5 - \left(1-P^{2/3}\right)^{1.21}\Bigg/ 4\left[(1-s)\frac{(3-5P)(1-P)}{2(3-5P)(1-2v_0)+3P(1+v_0)}+s\frac{(1-P)}{3(1-v_0)}\right] \qquad (2)$$

where $s = \dfrac{1}{1+e^{-100(P-0.4)}}$, P is the porosity, v_0 is Poisson's ratio for porous-free material. The Poisson's ratios for pore-free alumina, silica, and mullite are 0.23, 0.17, and 0.28 respectively.[22] For alumina-silica composite, this number was obtained by linear interpolation based upon the compositions.

RESULTS AND DISCUSSION

Phase Development

The phase development with the sintering temperature increase is shown in Fig. 1 for the 11.28 vol% alumina concentration sample. Since the silica nanoparticles are amorphous, the XRD pattern at room temperature displays the features of alumina (typical 2θ at 37.7°, 45.9°, and 66.9°) and sodium chloride (typical 2θ at 31.7°). With the increase of the sintering temperature, alumina and silica nanoparticles interact with each other, and inter-diffusion leads to new peak formation. At 1200°C, the new peaks are not developed well and the alumina and silica nanoparticles are still in the middle of the phase transformation. However, the new peaks are consistent with those of mullite. Upon further increase of the sintering temperature, the peaks in the XRD pattern gradually evolve and the mullite phase is identified at 1300°C (typical 2θ at 26.3°, 35.3°, 41.0° and 60.8°). It should be noticed that this phase transformation temperature for mullite is lower than the typical mullite formation temperature of 1400°C. This can be attributed to the small silica and alumina particle sizes that reduce the diffusion distance of the aluminum and silicon species and thermodynamically facilitate the phase transformation. However, residual alumina and silica phases (α-cristobalite, typical 2θ at 22.0°) still exist in the sintered material, indicating that the phase transformation is still in progress.

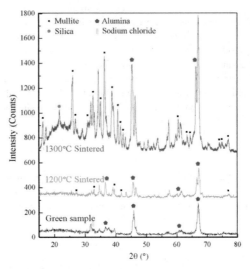

Fig. 1. XRD patterns for the 18 vol% solids loading composite with 11.28 vol% alumina concentration at different sintering temperatures.

Composition Effect on Porous Microstructure

The interaction between alumina and silica nanoparticles is confirmed by the SEM images. Shown in Fig. 2 are the microstructures of the 1300°C sintered samples with alumina concentration at 9.51 vol% , 11.28 vol% , and 11.80 vol%, respectively. The low magnification images show the porous microstructures while the inserted images taken at a high magnification display the grain level details.

Comparing the porous microstructures of the three different compositions, the composite with initial composition the same as stoichiometric mullite ($3Al_2O_3 \cdot 2SiO_2$) possesses the finest and most homogeneous microstructure. Specifically, the 9.51 vol% and the 11.80 vol% alumina samples have much larger pores compared to the 11.28 vol% alumina sample. The pore walls in either the 9.51 vol% or the 11.80 vol% alumina sample are rounded at the corners while the 11.28 vol% alumina sample displays sharper corners in the pore walls. This porous structure difference is caused by sintering of different composition samples, as the freeze-casting process mostly depends on physical interaction between the particles and the solidification front. At 1300°C temperature, excessive silica nanoparticles result in more extensive sintering with denser structures for the solid phase (pore walls). For the alumina excessive sample (11.80 vol% alumina), the pore walls have a coarser microstructure than the 11.28

vol% sample. At the grain level, alumina and silica nanoparticles have diffused together. However, the grain growth is fairly limited and individual silica and alumina grains are likely present.

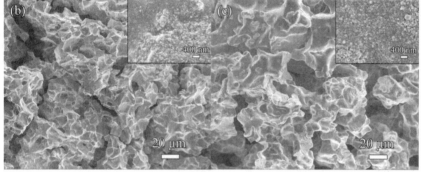

Fig. 2. Microstructure of the freeze-cast and 1300°C sintered composite with different compositions: (a) 9.51 vol% alumina, (b) 11.28 vol% alumina, and (c) 11.80 vol% alumina.

Composition Effect on Porosity

As shown in Fig. 3, the porosity before and after sintering at 1300°C for the 9.51 vol%, 11.28 vol%, and 11.80 vol% alumina samples are measured and plotted. The freeze-casting technique experiences a near net-shape process that the volume expansions are negligible. As a result, all of the green samples have a porosity of 82%. However, the porosity for the sintered samples varies according to the different compositions. The 9.51 vol% alumina sample shows the highest volume shrinkage and the sintered porosity is 59.6%, followed by the 11.80 vol% alumina sample which has a porosity of 69.0%. The 11.28 vol% alumina sample experiences the lowest volume shrinkage and possesses a porosity of 69.4%. Since the 11.28 vol% and 11.80 vol% alumina samples have very similar compositions, they have almost the same porosity. This porosity difference with the sample composition is consistent with the microstructure evolution. The excessive silica nanoparticles for the 9.51 vol% alumina sample lead to more extensive sintering and result in more densification compared to the stoichiometric mullite composition and the excessive alumina composition. The high porosity for the 11.28 vol% alumina sample is conserved after sintering. For the 11.80 vol% alumina sample, the sintering is also inhibited because of the higher sintering temperature nature of alumina nanoparticles.

It should also be mentioned that the porosities of the samples are highly interconnected. Although the absolute densities of the sintered samples can hardly be determined because the phase transformations are still in processing and the samples are mixtures of mullite, silica and alumina, the values should range from 2.80 g/cm^3 (the density of mullite) to ~3.14 g/cm^3 (the density of the silica-alumina composite based on interpolation of the components). The relative density (density containing closed pores) measured by a pycnometer (AccuPyc 1330, Micromeritics, Norcross, GA) is ~3.08 g/cm^3. Thus, the closed porosity is smaller than 2%, if any.

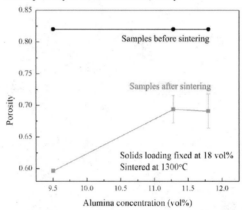

Fig. 3. Porosity of silica-alumina composite at different compositions.

Flexural Strength

Since the sample with stoichiometric mullite composition possesses the highest porosity and finest microstructure, the flexural strength test is taken on this sample. The strengthening effect upon sintering temperature increase is shown in Fig. 4. It should be noticed that the freeze-cast green sample is extremely fragile and the flexural strength is below 0.05 MPa, which cannot be experimentally measured (The sample is broken by the load ring before any test can be done). Upon sintering, the strength of the sample gradually improves and reaches 1.79 MPa at 1300°C. With the temperature increase, the aluminum and silicon species diffusion is intensified and the bonding between alumina nanoparticles and silica nanoparticles is strengthened. This bonding between particles/grains offers the sample extra strength as the applied load can be more effectively distributed throughout the sample. When the sintering temperature is increased to the one that mullite forms (above 1300°C), the sample strength is further improved as the mullite phase provides higher strength than either pure alumina phase or pure silica phase.

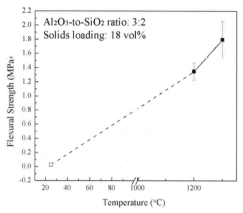

Fig. 4. Flexural strength of the 18 vol% solids loading composite with 11.28 vol% alumina concentration at different sintering temperatures.

CONCLUSIONS

Porous mullite with a high porosity of 69% is fabricated by freeze-casting and sintering alumina and silica nanoparticles. The mullite phase forms at a lower temperature (1300°C) compared to the traditional phase transformation temperature of 1400°C as the nano-sized particles provide enhanced phase transformation driving force. The sample with initial composition the same as the stoichiometric mullite possesses the finest microstructure and highest porosity compared to either alumina-rich composition or silica-rich composition. Excessive silica nanoparticles promote growth between particles/grains and porosity loss. With the increase of the sintering temperature, the samples become stronger as the bonding between particles/grains promotes the distribution of the applied load. At sintering temperature above 1300°C, mullite forms and provides the porous material with further improved strength.

ACKNOWLEDGEMENT

The authors acknowledge the financial support from National Science Foundation under grant No. CBET-0827246 and the American Chemical Society Petroleum Research Fund under grant No. 47421-AC9. Assistance from the Nanoscale Characterization and Fabrication Laboratory of Virginia Tech is greatly acknowledged.

REFERENCES

[1] M. Nakata, K. Tanihata, S. Yamaguchi, and K. Suganuma, Fabrication of Porous Alumina Sintered Bodies by a Gelate-Freezing Method, *J. Ceram. Soc. Jpn*, **90** [7] 2276–2279 (2007).
[2] S. Ding, Y. Zeng, and D. Jiang, Fabrication of Mullite Ceramics with Ultrahigh Porosity by Gel Freeze Drying, *J. Am. Ceram. Soc.,* **90** [7] 2276–2279 (2007).
[3] F. Ye, J. Zhang, H. Zhang, and L. Liu, Effect of Sintering Temperature on Microstructure and Mechanical Properties of Highly Porous Silicon Nitride Ceramics Produced by Freeze Casting, *Mater.*

Sci. Eng. A, **527** 6501-6504 (2010).

[4] W. Li, K. Lu, and J. Y. Walz, Formation, Structure and Properties of Freeze-Cast Kaolinite-Silica Nanocomposites, *J. Am. Ceram. Soc.,* **94** [4] 1256-1264 (2011).

[5] R. G. Chandran, B. K. Chandrashekar, C. Ganguly, and K. C. Patil, Sintering and Microstructural Investigation on Combustion Processed Mullite, *J. Eur. Ceram. Soc.,* **16** 843-849 (1996).

[6] X. Miao, Porous Mullite Ceramics from Natural Topaz, *Mater. Lett.,* **38** [2] 167-172 (1999).

[7] M. Iida, T. Sasaki, and M. Watanabe, Titanium Dioxide Hollow Microspheres with an Extremely Thin Shell, *Chem. Mater.,* **10** [12] 3780-3782 (1998).

[8] C. Tallon, D. Jach, R. Moreno, M. Nieto, G. Rokicki, and M. Szafran, Gelcasting of Alumina Suspensions Containing Nanoparticles with Glycerol Monoacrylate, *J. Eur. Ceram. Soc.,* **29** 875–880 (2009).

[9] Y. Liu, X. Liu, G. Li, and G. Meng, Low Cost Porous Mullite-Corundum Ceramics by Gelcasting, *J. Mater. Sci.,* **36** 3687-3692 (2001).

[10] K. Nakanishi, Pore Structure Control of Silica Gels Based on Phase Separation, *J. Mater. Sci.,* **4** 67-112 (1997).

[11] S. Deville, Freeze-casting of Porous Ceramics: A Review of Current Achievements and Issues, *Adv. Eng. Mater,* **10**[3] 155-69 (2008).

[12] L. Han, Z. Xu, Y. Cao, Y, Wei, and H. Xu, Preparation, Characterization and Permeation Property Al_2O_3, Al_2O_3-SiO_2 and Al_2O_3-Kaolin Hollow Fiber Membranes, *J. Membr. Sci.,* **372** 154-164 (2011).

[13] I. Aksay and J. A. Pask, Stable and Metastable Equilibria in the System SiO_2-Al_2O_3, *J. Am. Ceram. Soc.,* **58** [11-12] 507-512 (1975).

[14] Z. Z. Fang and H. Wang, Densification and Grain Growth during Sintering of Nanosized Particles, *Int. Mater. Rev.,* **53**[6] 326-52 (2008).

[15] K. Lu, Sintering of Nanoceramics, *Int. Mater. Rev.,* **53**[1] 21-38 (2008).

[16] G. L. Lecomte-Nana, J. P. Bonnet, and P. Blanchart, Investigation of the Sintering Mechanisms of Kaolin-Muscovite, *Appl. Clay Sci.,* **51** 445-451 (2011).

[17] T. Yang, H. B. Ji, S. Y. Yoon, B. K. Kim, and H. C. Park, Porous Mullite Composite with Controlled Pore Structure Processed Using a Freeze Casting of TBA-Based Coal Sly Ash Slurries, *Resour. Conserv. Recy.,* **54** 816-820 (2010).

[18] M. Tourbin and C. Frances, A Survey of Complementary Methods for the Characterization of Dense Colloidal Silica, *Part Part Syst Char,* **24**, 411-23 (2007).

[19] M. Tourbin and C. Frances, Monitoring of the Aggregation Process of Dense Colloidal Silica Suspensions in a Stirred Tank by Acoustic Spectroscopy, *Powder Technol,* **190**, 25-30 (2009).

[20] ASTM Designation C1499–04 American Society for Testing and Materials International, West Conshocken, PA (2004).

[21] M. Arnold, A. R. Boccaccini, and G. Ondracek, Prediction of the Poisson's Ratio of Porous Materials, *J Mater Sci,* **31**, 1643-46 (1996).

[22] H. Ledbetter, S. Kim, and D. Balzar, Elastic Properties of Mullite, *J. Am. Ceram. Soc.,* **81** [4] 1025–28 (1998).

LOW TEMPERATURE SINTERING OF GADOLINIUM-DOPED CERIA FOR SOLID OXIDE FUEL CELLS

Pasquale F. Lavorato, Leon L. Shaw
Department of Chemical, Materials, and Biomolecular Engineering
University of Connecticut, Storrs, CT 06269

ABSTRACT

Gadolinia-doped ceria (GDC, $Ce_{0.8}Gd_{0.2}O_{1.9}$) powder with particle sizes smaller than 10 nm is synthesized via co-precipitation. The effects of powder pre-treatment before sintering on sinterability of the GDC nanopowder are investigated. It is found that low speed, dry milling of the as-synthesized powder leads to enhancements in sinterability for sintering temperatures at 1000, 1100 and 1200°C. As a result, the pellets made from the dry milled powder exhibit the highest densities among all the sintered pellets investigated in this study. In contrast, calcination at 400°C of the as-synthesized powder leads to lower density pellets sintered at 1000 and 1100°C in comparison with those pellets made from the as-synthesized powder and the dry milled powder. The pellets using the dry milled powder sintered at 1200°C reach the density of 94% of the theoretical and contain only isolated micrometer and nanometer pores. As such, these GDC membranes can be used as the electrolyte for solid oxide fuel cell applications because they are impermeable to gaseous fuels and oxidants. The mechanisms for the enhanced sinterability due to dry milling and the decreased sinterability due to calcination are discussed.

INTRODUCTION

Gadolinia-doped ceria (GDC)-based fluorite cubic materials have attracted extensive attention worldwide lately owing to their potential as the electrolyte for solid oxide fuel cells (SOFCs) to operate at intermediate temperatures (IT) [1-11]. Such enormous interest derives from the much higher ionic conductivity of GDC electrolytes than the state-of-the-art electrolyte made of yttria-stablized zirconia (YSZ) [1-3]. There are multiple advantages for SOFCs to operate at IT (500 – 700°C). These include reducing the cost of the cell fabrication and materials in addition to the improved reliability and operational life [12,13]. However, operating temperatures at 500–700°C lead to high concentration polarization and activation polarization loss of electrodes, and require electrolytes to have high oxygen ion conductivities at these operation temperatures. Therefore, GDC with a higher ionic conductivity than that of YSZ at the temperature range of 300–700°C [1-3] has been studied extensively for electrolyte applications in IT-SOFCs [3-5].

Dense GDC membranes are needed for electrolyte applications. However, it is very difficult to achieve dense GDC bodies by conventional sintering at relatively low temperatures (< 1300°C) [4,7,8,14,15]. High sintering temperatures have to be used, but may destroy or adversely affect the other materials in SOFCs. Thus, lowering the sintering temperature of GDC electrolytes, which could lead to the possibility of co-sintering of the electrolyte and electrodes, is highly desirable.

Many studies have been conducted to reduce the sintering temperature of GDC [4,7,8,14,15]. Conventional methods of preparing dense ceria-based membranes typically entail two major steps. The first step is to synthesize ceria-based powders through several methods such as co-precipitation [7,8,16], solid-state reaction [17], mechanochemical processing [18], combustion synthesis [4,19], self-propagating synthesis [20], sol-gel process [10,11,21-23], sol-gel combustion approach [15], and hydrothermal synthesis [24]. The second step is to sinter the powder compact to form dense membranes [4,7,8,14,15,22,23]. Sintering temperatures as high as 1400–1600°C have been frequently used in order to achieve high density membranes (> 95% of the theoretical density) [4,8,15,23,25].

In this study, we have investigated the effect of powder treatment on the sintering temperature of GDC with the objective to co-sinter SOFCs using GDC as the electrolyte in the future. Since fine particle size is one of the key parameters for low temperature sintering [8], the synthesis condition of co-precipitation has been adjusted to produce ultrafine powder with particle sizes smaller than 10 nm. The powder attained is then subjected to different powder treatments with a goal to densify GDC at 1200°C. It is found that proper powder treatment before sintering of the GDC nanopowder is critical for densification, and GDC bodies with a density of 94% of the theoretical can be obtained via sintering at 1200°C when proper powder treatment is conducted. The details of our findings are reported below.

EXPERIMENTAL PROCEDURE

GDC Nanopowder Synthesis
The nanopowder of ceria doped with 20% gadolinium ($Ce_{0.8}Gd_{0.2}O_{1.9}$) was synthesized via a co-precipitation method [8]. Cerium nitrate, $Ce(NO_3)_3 \cdot xH_2O$ (Aldrich, 99.99%), and gadolinium nitrate, $Gd(NO_3)_3 \cdot xH_2O$ (Aldrich, 99.9%) were used in proper proportions to produce $Ce_{0.8}Gd_{0.2}O_{1.9}$. Ammonium hydroxide (NH_4OH) was used as the precipitant. The nitrates were dissolved in deionized water to make a 0.3 molar solution. The solution was added at room temperature dropwise to the NH_4OH precipitant, while stirred using a magnetic bar, and maintained at a pH of 10. The solution was then heated to 75°C while being stirred continuously and homogenized at that temperature for 1 h. The reaction produced a precipitate which was washed with distilled water and ethanol to a pH of 7, and then dried at 50°C in an oven.
The powder obtained was divided into three groups: the as-synthesized, the calcined, and the low speed dry milled. The calcined powder was attained by calcining the as-synthesized powder in air at 400°C for 4 hours, whereas a rolling mill was employed to prepare the dry milled powder. Low speed dry milling was conducted by loading the as-synthesized powder with a proper amount of steel balls into a polyethylene bottle and then milling at room temperature for 24 hours. These different powder pre-treatments before sintering could alter the powder characteristics, thereby impacting the sintering properties of GDC. Powder from each group was uniaxially pressed into pellets under a pressure of 300 MPa using steel dies of either 12.7 mm or 7.0 mm in diameter.

Sintering and Characterization of GDC Bodies
A thermomechanical analyzer (TMA, TA Instruments Q400) with a displacement resolution better than 0.5 nm was used to determine the dimensional change taking place during the sintering process. The 7-mm diameter pellets, from each powder group, were used for this purpose. In every TMA run, the TMA probe with a force of 0.05 N was maintained on the pellet placed on the TMA stage. The temperature of the TMA stage was controlled by lowering a mini-cylindrical furnace into place and subsequently adjusting the temperature of the furnace. The temperature profile for all the tests consisted of allowing the furnace to equilibrate at 50°C, followed by ramping up at 3°C/min to 900°C (which is the upper limit of the TMA instrument). A flowing air atmosphere with a flow rate of 50 mL/min was maintained in the entire holding and heating process.
The 12.7-mm diameter pellets were used for isothermal sintering studies at temperatures of 900°C 1000°C, 1100°C, and 1200°C with an air atmosphere. Sintering of samples at 900°C, 1000°C and 1100°C was conducted in a box furnace with heating and cooling rates of 3°C/min. Sintering of samples at 1200°C was conducted in an Al_2O_3 tube furnace under the same heating and cooling conditions.
The weight loss of the as-synthesized sample during the heating process was investigated via the thermogravimetric analysis (TGA) using a TA instrument (TGA Q500). The analysis was

performed with the following conditions. The sample of interest was loaded into a Pt-microbalance pan. The TGA system was then equilibrated at 60°C before heating to 1000°C with a constant heating rate of 20°C/min. The flow rate of air was maintained at 60 ml/min for the entire heating processes.

X-ray diffraction (XRD) analysis was performed for the as-synthesized, calcined and milled powders as well as the sintered pellets using a Druker, D8 Advanced diffractometer. The operation conditions for the XRD data collection were CuK$_\alpha$ radiation, 40 kV, 40 mA, 5°/min, and 0.02°/step. In addition to phase identification, XRD was also used to estimate the crystallite sizes of nanopowders and sintered bodies with the aid of the Scherrer formula without the consideration of internal strains [26].

$$D = \frac{0.9\lambda}{B_g(2\theta).\cos\theta} \qquad (1)$$

where $B_g(2\theta)$ is the broadening of the diffraction line measured at half maximum intensity, λ is the wave length of the X-ray radiation, and θ is the Bragg angle. The correction for instrumental broadening was conducted using the procedure described in [27] with the aid of coarse-grained silicon (Si) of 99.9% purity and the following equation,

$$B_g^{\,2}(2\theta) = B_h^{\,2}(2\theta) - B_f^{\,2}(2\theta) \qquad (2)$$

where B_g is the full width half maximum from the desired curve if there were no instrumental broadening, B_h is the full width half maximum from the sample, and B_f is the full width half maximum from the silicon standard.

The particle size and morphology of the as-synthesized GDC powder were analyzed using transmission electron microscopy (TEM, FEI Tecnai T-12). The fracture surfaces of the sintered pellets were used to investigate porosity and grain size. The imaging of fracture surfaces was conducted using a field-emission scanning electron microscope (FESEM, JEOL JSM 6335F).

The green and sintered densities of all samples were determined using the mass and volume of GDC pellets before and after sintering. The Archimedes method was also conducted for the sintered pellets. The theoretical density of GDC used to determine the percent density was 7.24 g/cm^3. This theoretical density was computed using the following formula [28].

$$\rho_{th} = \left(\frac{4\left[(1-x)M_{Ce} + xM_{Gd} + (2-0.5x)M_O\right]}{N_A a^3} \right) \qquad (3)$$

where M is the atomic mass, x the dopant fraction (0.2 in this case), a the lattice parameter (5.423 nm) and N_A the Avogadro's number (6.022 x 10^{23}).

RESULTS AND DISCUSSION

Characterization of the As-Synthesiz ed GDC: Fig. 1 shows the XRD spectrum of the as-synthesized GDC powder. It is found that the as-synthesized GDC powder matches well to the phase pure $Ce_{0.8}Gd_{0.2}O_{1.9}$ (JCPDF No. 75-0162). With the aid of Scherrer formula [26] and using the full width at the half maximum (FWHM) of the 220 peak, the crystallite size of GDC is estimated to be 9 nm. This crystallite size is found to be consistent with the particle size observed using TEM (Fig. 2).

Thus, it can be concluded that the GDC powder generated from the co-precipitation method is composed of single crystals with an average size of 9 nm in diameter.

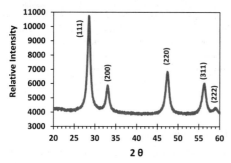
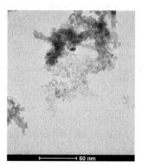

Fig. 1 XRD spectrum of the as-synthesized GDC.   **Fig. 2** TEM image of the as-synthesized GDC.

Powder Characteristics after the Pre-Treatment before Sintering: Fig. 3 compares XRD spectra of the as-synthesized powder and the powder after calcination at 400°C. The XRD spectrum of samples sintered at 1200°C for 2 h is also included for comparison. It can be seen that the powder remains to be phase pure $Ce_{0.8}Gd_{0.2}O_{1.9}$ after the calcination pre-treatment. Using the Scherrer formula [26] and the FWHM of the 220 peak, it is found that calcination at 400°C does not induce crystal growth because the crystallite size remains to be 9 nm. Low speed, dry milling at room temperature does not change the crystallite size either (not shown here). Furthermore, there is no new phase formation due to dry milling at room temperature. These observations are consistent with our expectation because the two main functions of the low speed, dry ball milling are (i) reducing the sizes of micrometer or sub-micrometer agglomerates and (ii) inducing the sliding of nanoparticles to pass each other when nanoparticles are trapped during ball-to-ball or ball-to-wall collision. Note that even with high-energy ball milling (i.e., high speed, dry ball milling), the stress generated during ball-to-ball collision is not high enough to decrease particle sizes below 10 nm for many materials [29-33].

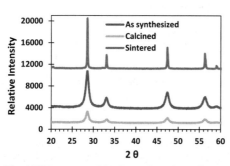

Fig. 3 XRD spectra of GDC powders after different pre-treatments before sintering. The XRD spectrum of GDC samples sintered at 1200°C is also included for comparison.

Sintering Properties of Various Powders: Table 1 summarizes the green densities and sintered densities of various GDC pellets with different powder pre-treatments. It is noted that the pre-treatment, either calcination or low speed milling, does not change the density of green pellets. Thus, the green density is mainly controlled by the pressure (300 MPa) used in the uniaxial pressing. However, it is interesting that the densities of sintered pellets exhibit a clear dependence on the pre-treatment method. For example, taking the GDC pellets sintered at 1200°C for consideration, the GDC pellet with the pre-treatment of low speed, dry milling has the highest density among the three powder conditions studied. Furthermore, achieving a density of 94% of the theoretical at 1200°C is better than most of the studies conducted previously [4,8,15,23,25]. If the densities of sintered bodies at other temperatures are compared, the pellets with low speed, dry milled powder also exhibit the highest density among the three powder conditions. Therefore, it can be concluded that low speed, dry milling is beneficial for attaining sintered bodies with high densities.

Fig. 4 shows the optical images of the polished surfaces of GDC pellets, sintered using the low speed, dry milled powder at 1000, 1100 and 1200°C for 2 h. Note that optical imaging is a good method to assess the presence of micrometer-sized pores and their distribution within the sample, but not suitable for detection of nanometer-sized pores. With this in mind, one can conclude that a few isolated large pores of less than 10 μm in diameter are present in the pellets sintered at 1100°C (Fig. 4b) and 1200°C (Fig. 4c). Further, these pores distribute quite uniformly throughout the samples. However, pores of less than 20 μm in diameter are still present and distribute un-uniformly in the pellets sintered at 1000°C (Fig. 4a). This trend of the size of micrometer pores is consistent with the trend revealed from the density data determined via weight and volume measurements (Table 1). Quantitatively, however, the presence of a small number of micrometer pores cannot account for the relatively low density of the pellets sintered at 1000°C (only 68.2% dense). Thus, it is highly likely that nanometer pores are present in the pellets sintered at 1000°C and possibly in the pellets sintered at 1100°C. Such inference is confirmed by SEM observations, as discussed below.

Table 1. Densities of GDC pellets at the green state and after sintering

Powder Treatment	Green Density	Density Sintered at 900°C, 2h	Density Sintered at 1000°C, 2h	Density Sintered at 1100°C, 2h	Density Sintered at 1200°C, 2h
As-synthesized	45.6%	53.4%	64.3%	88.2%	87.7%
Calcined @ 400°C, 4 h	45.7%	53.7%	58.8%	84.2%	91.9%
Low speed, dry milled for 24 h	45.0%	-	68.2%	90.7%	93.7%

* Density was determined via weight and volume measurements.

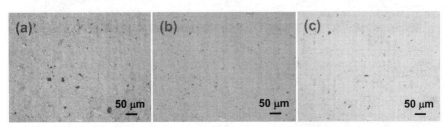

Fig. 4 Optical images of the polished surfaces of GDC pellets, using the dry milled powder sintered at (a) 1000°C for 2 h, (b) at 1100°C for 2 h, and (c) at 1200°C for 2 h.

Fig. 5 shows the SEM images of the fracture surfaces of GDC pellets prepared under several different conditions. It is clear that pellets sintered at 1000°C contain many nanometer pores because densification is very limited at this temperature. However, sintering at 1100°C has resulted in substantial densification even though many nanometer pores are still present at this temperature. Nevertheless, the presence of these nanometer pores can account for the low densities determined from the weight and dimensions of the pellets (e.g., 68.2% and 90.7% dense for the pellets sintered using the dry milled powder at 1000 (Fig. 5a) and 1100°C (Fig. 5c), respectively). These observations are also applicable to the pellets sintered using the calcined powder (not shown here). Thus, combining Figs. 4 and 5, one can conclude that all pellets sintered at 1000 and 1100°C contain a large number of nanometer pores with only very limited numbers of micrometer pores.

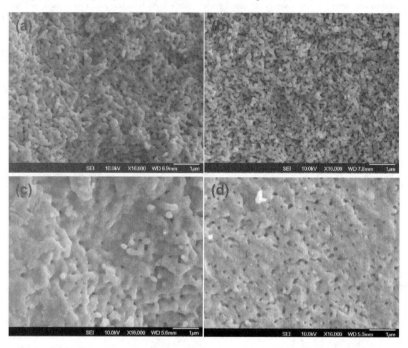

Fig. 5 SEM images of the fracture surfaces of various pellets: (a) sintered using the dry milled powder at 1000°C for 2 h, (b) sintered using the as-synthesized powder at 1000°C for 2 h, (c) sintered using the dry milled powder at 1100°C for 2 h, and (d) sintered using the as-synthesized powder at 1100°C for 2 h.

The situation for the pellets sintered at 1200°C appears to be quite different from that of the pellets sintered at 1000 and 1100°C. Fig. 6 shows the fracture surfaces of various pellets sintered at 1200°C. A comparison between Figs. 5 and 6 reveals that nanometer pores have been reduced drastically after sintering at 1200°C, indicating that most of nanometer pores can be effectively eliminated at 1200°C. In particular, the pellets made from the dry milled powder appear to have a density at 98% of the theoretical or higher (Fig. 6c) if only nanometer pores are considered. The discrepancy between the SEM observation of the fracture surface (Fig. 6c) and the density measured from the weight and dimensions (93.7% in Table 1) is due to the presence of a few micrometer pores as shown in Fig. 4c. The discrepancies for the pellets prepared from the as-synthesized (Fig. 6a and

87.7% in Table 1) and calcined powders (Fig. 6b and 91.9% in Table 1) can also be attributed to the presence of a few micrometer pores in these samples (not shown here). Based on these observations, one can conclude that all pellets sintered at 1200°C contain isolated nanometer and micrometer pores. For the pellets sintered using the calcined or dry milled powders at 1200°C, one can further conclude that these pellets can be used as the electrolyte membranes for SOFC applications because they contain only isolated pores and are thus impermeable to gaseous fuels and oxidants used in SOFCs.

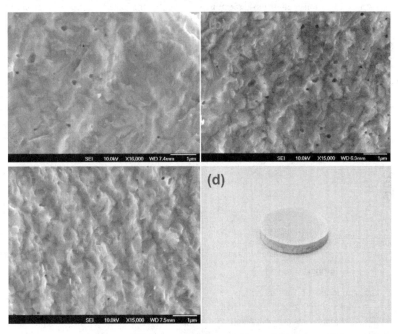

Fig. 6 SEM images of the fracture surfaces of various pellets: (a) sintered using the as-synthesized powder at 1200°C for 2 h, (b) sintered using the calcined powder at 1200°C for 2 h, and (c) sintered using the dry milled powder at 1200°C for 2 h. (d) A photo of a pellet sintered using the dry milled powder at 1200°C for 2 h, indicating the mechanical integrity of the sintered body.

The discussion above based on the optical and SEM observations as well as the density measurement unambiguously indicates that the dry milled powder has the best sinterability among all the powder conditions investigated in this study. It should be emphasized that the pellets sintered from the dry milled powder remain to be the phase pure GDC, as shown in Fig. 3. The difference between the powder and sintered bodies is the average crystallite size estimated from the peak broadening, which has increased from 9 nm before sintering to 45 nm after sintering at 1200°C. This observation also holds for the pellets sintered from the as-synthesized and the calcined powders. Thus, we have attributed the best sinterability of the dry milled powder to the function of ball milling in reducing the sizes of micrometer or sub-micrometer agglomerates. Decreasing the sizes of micrometer or sub-micrometer agglomerates will result in more uniform particle packing within green bodies, which in turn can promote densification. It is well established that non-uniform particle packing leads to preferential densification of localized regions, which results in microstructural defects and prevents complete densification during sintering [34,35]. Dry milling eliminates large agglomerates and thus promotes more uniform particle packing, thereby enhancing densification.

To understand why calcination at 400°C leads to a lower sinterability than the as-synthesized powder at 1000 and 1100°C, we have conducted TMA and TGA experiments to determine the dimensional change and weight loss behaviors during the sintering process. Shown in Fig. 7 are the shrinkage strains of various pellets during heating up to 900°C. It is noted that the pellet made from the dry milled powder exhibits the largest shrinkage at 900°C. This phenomenon is consistent with the fact that the pellets made from the dry milled powder have the highest densities after sintering at 1000, 1100 and 1200°C (Table 1). The shrinkage strain for the pellet made from the calcined powder is also consistent with the density measurement, i.e., the calcined powder leads to lower sintered densities than the as-synthesized powder at 1000 and 1100°C. However, it is noted that the shrinkage behavior is similar for the pellets made of the as-synthesized and dry milled powders below 400°C. In contrast, the pellet made of the calcined powder exhibits little shrinkage below 400°C. The different behavior displayed by the pellet made of the calcined powder is due to the calcination at 400°C before sintering, as discussed below.

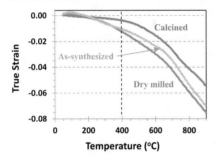

Fig. 7 The true strain of various pellets vs temperature determined using the TMA instrument. All experiments are conducted in air with a heating rate of 3°C/min.

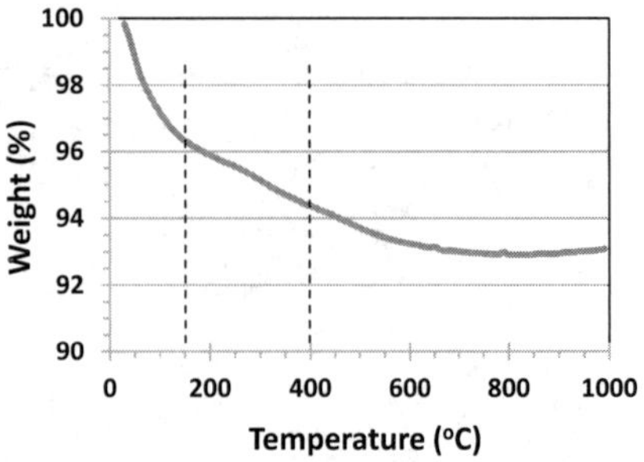

Fig. 8 Thermogravimetric analysis of the as-synthesized powder with a heating rate of 20°C/min and a flowing air atmosphere.

Fig. 8 shows that calcination at 400°C can result in weight loss of ~ 6%. Based on the weight loss rate (indicated by the slope), we propose that the weight loss below 400°C is mainly due to two processes. First, the weight loss from room temperature to ~150°C is related to the evaporation of physisorbed H_2O and O_2 on the surface of the GDC powder. This desorption results in about 4% weight loss. Second, the weight loss from ~150 to 400°C is mainly due to the decomposition of hydroxides on the surface of the powder because the surface of GDC particles is likely covered with a thin hydrated layer given the fact that the GDC powder is made through co-precipitation in water. The proposed weight loss mechanisms are consistent with the previous studies, showing that the desorption of physisorbed H_2O from CeO_2 typically occurs at temperatures up to 150°C [36], while complete dehydration takes place between 350 and 470°C, depending on the detailed chemistry of doped CeO_2 [36,37]. By comparing Figs. 7 and 8, one can infer that dehydration may result in shrinkage of GDC particles. As a result, the pellet made from the calcined powder exhibits little shrinkage below 400°C since the calcined particles have already gone through shrinkage during calcination at 400°C. The lack of shrinkage below 400°C leads to lower densities for the pellets made of the calcined powder sintered at 1000 and 1100°C (Table 1), suggesting that calcination at 400°C is not beneficial for attaining high densities with low temperature sintering. Sintering at 1200°C, however, results in a high density for the calcined powder than the as-synthesized powder. The mechanism for this phenomenon is not clear yet and remains to be studied in the future.

CONCLUSION

GDC ($Ce_{0.8}Gd_{0.2}O_{1.9}$) powder with particle sizes smaller than 10 nm is synthesized via the co-precipitation method. The effects of powder treatment before sintering on sinterability of the GDC nanopowder are investigated. It is found that low speed, dry milling of the as-synthesized powder leads to enhancements in sinterability with sintering temperatures at 1000, 1100 and 1200°C. In contrast, calcination at 400°C of the as-synthesized powder leads to lower density pellets sintered at 1000 and 1100°C in comparison with those pellets made from the as-synthesized powder and the dry milled powder. The high sinterability of the dry milled powder has been attributed to the function of ball milling in reducing the size of micrometer or sub-micrometer agglomerates, which in turn leads to more uniform particle packing within green bodies, thereby promoting the subsequent densification. It is suggested that the reduced sinterability induced by calcination at 400°C is associated the shrinkage taking place during calcination. Such shrinkage before powder compaction results in little shrinkage in the subsequent sintering process below 400°C. As a result, the pellets made from the calcined powder exhibit the lowest densities among all pellets when sintering temperature is at 1000 or 1100°C. The dry milled powder can result in sintered bodies with the density of 94% of the theoretical. Further, these sintered membranes can be used as the electrolyte for SOFC applications because only isolated pores are present in these membranes.

REFERENCES

1. B. C. H. Steele, Appraisal of $Ce_{1-y}Gd_yO_{2-y/2}$ Electrolytes for IT-SOFC Operation at 500°C, *Solid State Ionics*, **129**, 95–110 (2000).
2. H. Inaba, H. Tagawa, Ceria-Based Solid Electrolytes, *Solid State Ionics*, **83**, 1-16 (1996).
3. K. Eguchi, T. Setoguchi, T. Inoue, H. Arai, Electrical Properties of Ceria-Based Oxides and Their Application to Solid Oxide Fuel Cells, *Solid State Ionics*, **52**, 165-172 (1992).
4. C. Xia, M. Liu, Microstructures, Conductivities, and Electrochemical Properties of $Ce_{0.9}Gd_{0.1}O_2$ and GDC-Ni Anodes for Low-Temperature SOFCs, *Solid State Ionics*, **152-153**, 423-430 (2002).
5. J. S. Ahn, S. Omar, H. Yoon, J. C. Nino, and E. D. Wachsman, Performance of Anode-Supported Solid Oxide Fuel Cell using Novel Ceria Electrolyte, *J. Power Sources*, **195**, 2131–2135 (2010).
6. B. Dalslet, P. Blennow, P. Hendriksen, N. Bonanos, D. Lybye, M. Mogensen, Assessment of Doped Ceria as Electrolyte, *J. Solid State Electrochem.*, **10**, 547-561 (2006).
7. C. Guizard, A. Julbe, O. Robbe and S. Sarrade, "Synthesis and Oxygen Transport Characteristics of Dense and Porous Cerium/Gadolinium Oxide Materials: Interest in Membrane Reactors," *Catalysis Today*, **104**, 120-125 (2005).
8. R. S. Torrens, N. M. Sammes, G. A. Tompsett, Characterisation of $(CeO_2)_{0.8}(GdO_{1.5})_{0.2}$ Synthesized using Various Techniques, *Solid State Ionics*, **111**, 9-15 (1998).
9. J. M. Im, H. J. You, Y. S. Yoon, D. W. Shin, Synthesis of Nanocrystalline $Gd_{0.1}Ce_{0.9}O_{2-x}$ for IT-SOFC by Aerosol Flame Deposition, *Ceram. Int.*, **34**, 877-881 (2008).
10. L. Shaw, C. Shen, E. L. Thomas, Synthesis of Gadolinia-Doped Ceria Gels and Powders from Acetylacetonate Precursors, *J. Sol-Gel Sci. Technol.*, **53**, 1-11 (2010).
11. C. Shen, L. Shaw, FTIR Analysis of the Hydrolysis Rate in the Sol-Gel Formation of Gadolinia-Doped Ceria with Acetylacetonate Precursors, *J. Sol-Gel Sci. Technol.*, **53**, 571-577 (2010).

12. S. C. Singhal, Solid Oxide Fuel Cells for Stationary, Mobile, and Military Applications, *Solid State Ionics*, **152-153**, 405-410 (2002).
13. N. Q. Minh, Solid Oxide Fuel Cell Technology - Features and Applications, *Solid State Ionics*, **174**, 271-277 (2004).
14. M. A. Panhans, R. N. Blumenthal, A Thermodynamic and Electrical Conductivity Study of Nonstoichiometric Cerium Dioxide, *Solid State Ionics*, **60**, 279–298 (1993).
15. S. Pinol, M. Najib, D. M. Bastidas, A. Calleja, X. G. Capdevila, M. Segarra, F. Espiell, J. C. Ruiz-Morales, D. Marrero-Lopez, P. Nunez, Microstructure-Conductivity Relationship in Gd- and Sm-Doped Cerai-Based Electrolytes Prepared by the Acrylamide Sol-Gel-Related Method, *J. Solid State Electrochem.*, **8**, 650-654 (2004).
16. D. Terribile, A. Trovarelli, J. Llorca, C. de Leitenburg, G. Dolcetti, The Synthesis and Characterization of Mesoporous High-Surface Area Ceria Prepared using a Hybrid Organic/Inorganic Route, *J. Catal.* **178**, 299–308 (1998).
17. X. Yu, F. Li, X. Ye, X. Xin, Z. Xue, Synthesis of Cerium(IV) Oxide Ultrafine Particles by Solid-State Reactions, *J. Am. Ceram. Soc.*, **83**, 964–966 (2000).
18. T. Tsuzki, P. G. McCormik, Synthesis of Ultrafine Ceria Powders by Mechanochemical Processing. *J. Am. Ceram. Soc.*, **84**, 1453–1458 (2001).
19. S. B. Boskovic, B. Z. Matovic, M. D. Vlajic, V. D. Krstic, Modified Glycine Nitrate Procedure (MGNP) for the Synthesis of SOFC Nanopowders, *Ceram. Int.* **33**, 89–93 (2007).
20. S. Boskovic, D. Durovic, Z. Dohcevic-Mitrovic, Z. Popovic, M. Zinkevich, F. Aldinger, Self Propagating Room Temperature Synthesis of Nanopowders for SOFC, *J. Power Sources*, **145**, 237–242 (2005).
21. M. A. Thundathil, W. Lai, L. Noailles, B. S. Dunn, S. M. Haile, High Surface-Area Ceria Aerogel, *J. Am. Ceram. Soc.*, **87**, 1442-1445 (2004).
22. C. Xia, Y. Zhang, M. Liu, LSM-GDC Composite Cathodes Derived from a Sol-Gel Process, *Electrochem. Solid-State Lett.*, **6**, A290-A292 (2003).
23. W. Huang, P. Shuk, M. Greenblatt, Properties of Sol-Gel Prepared $Ce_{1-x}Sm_xO_{2-x/2}$ Solid Electrolytes, *Solid State Ionics*, **100**, 23-27 (1997).
24. X. Xu, Z. Jiang, X. Fan, C. Xia, LSM-SDC Electrodes Fabricated with an Ion-Impregnating Process for SOFCs with Doped Ceria Electrolytes, *Solid State Ionics*, **177**, 2113-2117 (2006).
25. M. Hirano, E. Kato, Hydrothermal Synthesis of Cerium (IV) Oxide, *J. Am. Ceram. Soc.*, **79**, 777-780 (1996).
26. L. Azaroff, *Elements of X-ray Crystallography*, McGraw-Hill, New York, 1968.
27. Z.-G. Yang, L. Shaw, Synthesis of Nanocrystalline SiC at Ambient Temperature through High Energy Reaction Milling, *Nanostruct. Mater.*, **7**, 873-886 (1996).
28. B. D. Cullity, S. R. Stock, *Elements of X-ray Diffraction*, 3rd Ed., Prentice Hall, Upper Saddle River, NJ, 2001
29. Z.-G. Yang, R.-M. Ren, L. Shaw, Evolution of Microstructures and Nitrogen Sorption during High Energy Milling of Si in Ammonia, *J. Am. Ceram. Soc.*, **83**, 1897-1904 (2000).
30. R.-M. Ren, Z.-G. Yang, L. Shaw, Polymorphic Transformation and Powder Characteristics of TiO_2 during High Energy Milling, *J. Mater. Sci.*, **35**, 6015-6026 (2000).
31. X. Wan, T. Markmaitree, W. Osborn, L. Shaw, Nanoengineering-Enabled Solid-State Hydrogen Uptake and Release in the $LiBH_4$ plus MgH_2 System, *J. Phys. Chem. C*, **112**, 18232-18243 (2008).
32. C. A. Galán, A. L. Ortiz, F. Guiberteau, L. Shaw, High-Energy Ball Milling of ZrB_2 in the Presence of Graphite, *J. Am. Ceram. Soc.*, **93**, 3072–3075 (2010).
33. C. A. Galán, A. L. Ortiz, F. Guiberteau, L. Shaw, Crystallite Size Refinement of ZrB_2 by High-Energy Ball Milling, *J. Am. Ceram. Soc.*, **92**, 3114–3117 (2009).
34. F. F. Lange, Sinterability of Agglomerated Powders, *J. Am. Ceram. Soc.*, **67**, 83-89 (1984).

35. A. G. Evans, Considerations of Inhomogeneity Effects in Sintering, *J. Am. Ceram. Soc.*, **65**, 497-501 (1982).
36. E. C. C. Souza, E. N. S. Muccillo, Effect of Solvent on Physical Properties of Samaria-Doped Ceria Prepared by Homogeneous Precipitation, *J. Alloys Comp.*, **473**, 560–566 (2009).
37. N. Audebrand, J.-P. Auffrédic, D. Louer, An X-ray Powder Diffraction Study of the Microstructure and Growth Kinetics of Nanoscale Crystallites Obtained from Hydrated Cerium Oxides, *Chem. Mater.*, **12**, 1791-1799 (2000).

Nanotechnology for Energy, Healthcare, and Industry

CURRENT STATUS AND PROSPECTS OF NANOTECHNOLOGY IN ARAB STATES

Bassam Alfeeli, Ghada Al-Naqi*, and Abeer Al-Qattan*

Department of Advanced Systems, Kuwait Institute for Scientific Research
Safat, Kuwait
*These authors equally contributed to this work

ABSTRACT

A growing number of nanotechnology research, education and industry initiatives have been recently launched by several Arab States to quickly build scientific capacity and track the worldwide developments in nanotechnology. Some countries, namely the oil rich States, have allocated large funds to support these initiatives. This comprehensive commitment is intended to serve national interests in energy, water and food supply, medicine, and local industry. Other Arab States are also pursuing nanotechnology, however with fewer funds and more human resources. This study assesses current status and prospects of nanotechnology in the Arab Republic of Egypt, Kingdom of Saudi Arabia, State of Kuwait, State of Qatar, and United Arab Emirates. The study is aimed at having an overview of the status of existing, underdevelopment, and planed educational and research programs as well as commercial establishments relevant to nanotechnology. The overview also includes nanotechnology research focus areas, needs, challenges, and opportunities.

INTRODUCTION

Developed countries have long recognized the social and economic potentials of nanotechnology and quickly reacted to the rise of this technology by employing considerable resources to advance it and subsequently benefit from it. Being the world's first nation to launch national nanotechnology program in 2000, the United States invested $12.4 billion in the last decade on its national nanotechnology initiative[1]. Moreover, according to ISESCO[2], in the United States alone, there are more than 300 laboratories along with 700 companies working on the development of nanotechnology. Japan, where the term "nanotechnology" was coined, is allocating $800 million for nanotechnology development. China has over 300 organizations working on the development of various kinds of nanomaterials.

Developing countries such as the Arab States have recently realized their lag in taking part of the international effort to develop nanotechnology. In fact, Arab States are entering a new phase of political reform and economic development. With emphasis on science and technology (S&T) for promoting sustainable development, a growing number of nanotechnology research, education and industry initiatives have been recently launched by several Arab States to quickly build scientific capacity and track the worldwide developments in nanotechnology. This study assesses current status and prospects of nanotechnology in Arab States. The countries examined are Arab Republic of Egypt, Kingdom of Saudi Arabia, State of Kuwait, State of Qatar, and United Arab Emirates. The study seeks to document the approach of the different countries along four parameters: national commitment; infrastructure and funding; research; and education and training. The descriptive overview covers the status of existing, underdevelopment, and planed educational and research programs as well as commercial establishments relevant to nanotechnology. This study drew information from publically available official documents and reports as well as material collected from official websites of institutions discussed. All reproduced text in this article has been released into the public domain by its authors.

BACKGROUND

The Arab World consists of 22 countries covering about 10% of the world's land and is home to about 300 million representing 4.5% of the world's population. The UNESCO Science Report[3] groups the Arab States into three groups in terms of per capita income. The first group is characterized by almost total economic dependence on oil (Bahrain, Emirates, Kuwait, Oman, Qatar, and Saudi), with gross domestic product (GDP) per capita income being highest in Qatar and lowest in Oman. Around 11% of the Arab population belongs to this group of States. The higher education systems and science, technology and innovation (STI) in these States are new but developing rapidly thanks to sizeable investments by their heads of State and governments. The second group encompasses Algeria, Egypt, Iraq, Jordan, Lebanon, the Libyan, Morocco, Syria and Tunisia. Here, GDP per capita is highest in the Libyan and lowest in Egypt. Although the States in this category have modest oil reserves, with the notable exception of Iraq and Libyan, they possess relatively mature higher education infrastructure including some of the oldest universities in the Arab World. The population of this group amounts to around 70% of the population in the Arab World. The third group is characterized by limited or underdeveloped natural resources and an equally small supply of trained human resources. States in this group also possess some of the lowest GDP per capita in the world, which classifies them as least developed countries. They are Comoros, Djibouti, Mauritania, Sudan and Yemen. This group of countries represents around 19% of the total population of the Arab World.

According to Sawahel[4], on average, Arab States allocate less than 0.2% of their GDP on research compared to 1.6% in East Asia countries and 2.6% in developed countries. However, the budget allocated for purchasing armament in these States surpasses health, education, and research budgets combined. Nevertheless, in recent years the Arab leaders recognized the importance of education and STI. At the annual Arab summit and for the past 5 consecutive years (Sudan 2006, Saudi 2007, Syria 2008, Qatar 2009, and Libya 2010 summits) the Arab leaders adopted several decisions to establish science-based economy and knowledge society. One of the decisions was to increase the expenditure on research and development (R&D) to about 2.5% of GDP[5]. They also agreed to make education and scientific research permanent items at all future Arab summits. The Arab States don't only share common language and culture but also R&D priorities which include water and energy. The traditional sector of agriculture and the relatively new fields of information and communication technologies (ICTs), nanotechnology and biotechnology are also viewed as priority research areas as stated in the 20th Arab League summit held in Syria 2008.

But, there are a number of challenges that face the development of S&T in the Arab States including: (1) lack of supportive government policies; (2) limited and inconsistent R&D funding; (3) a small scientific research community; and (4) limited venture capital. In particular, there is no national commitment to establish organizations dedicated to devising strategies that address S&T. There is a need for policy-making body to formulate national policies that link R&D with national development priorities and execute technological development programs that fit the national economic and social objectives. Improving innovation requires a political decision and must be supported by a clear vision. Moreover, Arab States lack advance science-base and modern technology-base. There is a need to build up the infrastructure necessary for S&T. Constrains on R&D in Arab States are not only limited to the weak institutional structure but also inefficient administrative arrangement. There is also a need to develop human capacity. There are a small number of Arab scientists contributing to the scientific research output in the world. Their contribution account for only 1% of the world's scientific production[4]. Although this is gradually changing, the desire for scientific enquiry and activity in the Arab society is yet to be strengthened. For the case of nanotechnology development, the Arab States need to create a critical mass of researchers and scientist specializing in nanotechnology to effectively enhance their innovation capability[6].

CURRENT STATUS

Egypt

Egypt does not have formally enacted national science policy[7]. However, the number of well-established scientific institutions indicates the existence of an implicit national policy[8]. Moreover, in 2007 a presidential decree was issued to establish Science and Technology Development Fund (STDF) to support the scientific research and technological development. Another presidential decree was issued in the following year to establish the Higher Council for Science and Technology (HCST). The council is chaired by the Prime Minister and includes the minster of higher education and scientific research and a group of prominent Egyptian scientists. HCST was established to promote R&D in the country and identify priority areas which included health, water resources, renewable energy, food and agriculture, and space technology. It didn't explicitly state nanotechnology as a priority. In 2009, the so called Specialized Scientific Councils (SCCs) were established to provide scientific advices, studies and strategic plans on issues to the policy makers and the community[9].

As R&D continues to expand, efforts are being made to establish nanotechnology development in Egypt. Efforts to establish micro/nano research and fabrication facility started as early as 2003 with the establishment of Yousef Jameel Science and Technology Research Center (YJ-STRC) at the American University in Cairo (AUC). YJ-STRC was the fruit of the generous support ($8 million over 5 years) of a Saudi businessman and AUC alumnus Yousef Jameel. His vision was to create nanotechnology center of excellence at AUC. The center now houses class-100 clean room and state-of-the-art fabrication and characterization equipment. It should be noted that AUC offers Masters of Science in Nanotechnology and Doctor of Philosophy in Applied Sciences with specializations in Nanotechnology. To date, YJ-STRC has secured $13 million in funding and recruited 17 high profile faculty members with diverse backgrounds that includes electrical and mechanical engineering, material science, physics, chemistry, and biology. Furthermore, with the support of 8 postdoctoral fellows, 13 doctoral students, 20 master's students, and 4 technical staff, YJ-STRC conducts its work through 6 research groups: micro- and nano-systems, nanostructured materials, surface chemistry, biotechnology, environmental science and engineering, and novel diagnostics and therapeutics. Scientific production started as early as 2005 and to date, YJ-STRC holds 8 patents and published more than 70 articles in international journals and presented more than 65 papers at various international conferences. The research groups are serviced by well-equipped research facilities that include: micro- and nano-systems fabrication, materials synthesis, biotechnology, surface chemistry

YJ-STRC has also established ties with several international institutes and universities including: National Institute of Materials Science in Japan, Kyoto University in Japan, Royal institute of technology in Sweden, Universidade Nova de Lisboa in Portugal, Vrije Universiteit in the Netherlands, Stanford University, Virginia Tech, and Oregon Health & Science University in the U.S. Moreover, the center established a partnership agreement with King Abdullah University of Science and Technology (KAUST) in Saudi.

In 2006, the Nile University (NU) was established, by support from international and national companies represented by the Egyptian Foundation for Technological Education Development (EFTED), as a not-for-profit, privately-owned, and autonomously-managed university. NU is the first academic institution in Egypt to be founded by a partnership between the private sector (EFTED), government, business, and industry. NU was allocated around 0.5 km^2 of land and two buildings by the government. The buildings were equipped with technical equipment, information technology infrastructure and furniture paid for with more than $10 million raised by the university. NU has 6 faculty members working on nanotechnology related research and offers a Master of Science in Nanoscience and Technology degree. NU also houses two nanotechnology centers. The Center for Nanotechnology (CNT) which was establish based on collaboration efforts with Northwestern University in U.S. CNT researchers work on printed electronics, membrane technology, and renewable energy. The Nanoelectronics Integrated Systems Center (NISC) is funded by: Intel, Mentor Graphics, British Petroleum, European Union, Cypress Semiconductor Corp., Egyptian Information Technology Industry Development Agency (ITIDA), STDF, and National Telecom Regulatory Authority. NISC is pursuing research in areas that include: high performance integrated circuits (ICs), computer aided design ICs, low power circuit design, hardware for wireless sensor network, MEMs, and sensor and actuator design.

In an attempt to capture the currently underutilized human potentials in Egypt, IBM teamed up with ITIDA and STDF to create Egypt's first national research laboratory in 2008, the Egypt-IBM Nanotechnology Research Center (EGNC). The idea was to have IBM experts work with local scientists and engineers on advanced nanoscience and nanotechnology projects. With $30 million in seed money for the first three years, EGNC is focusing on solar energy technology and water desalination. The startup work force of EGNC was about 100 employees but expected to grow to up to 1,000 within the next few years. Current research areas include: thin-film silicon photovoltaics, spin-on carbon–based electrodes for thin film photovoltaics, energy recovery from concentrator photovoltaics for desalination, and computational modeling and simulation.

As for commercial establishments, three companies were formed by Egyptian researchers to serve the nanotechnology market in Egypt and the region. They mainly provide consultancy services. One of the companies, however, grew to provide nanomaterials characterization services and even to manufacture and sell nanomaterials. According to the company's website, their products include: metallic nanostructures, semiconductor nanostructures, magnetic nanostructures, hybrid nanostructures, and oxides. The company also runs several R&D projects related to water desalination, solar energy, and biomedical imaging.

Emirates

Although Emirates does not have formally enacted national science policy, it took several measures to promote R&D within the country. In 2010, the government released "UAE Vision 2021", a document outlining the vision for UAE in all fields until 2021 when the country will celebrate its 50th anniversary[10]. The document lists the promotion of innovation and R&D as one of the country's priorities. There is no specific mention of nanotechnology as a national priority. Nevertheless, it states that Emiratis will continue to attract the best talent from around the world and offer fulfilling employment and an attractive place to live to retain the finest and most productive workers and entrepreneurs. The plan also calls for knowledge-based, highly productive and competitive economy that will rival the best in the world by investing in science, technology, and R&D. According to the document, this goal will be achieved by supporting practical programs such as start-up incubators and cultivating a culture of risk-taking where hard work, boldness and innovation are rewarded.

In 2000, the ruler of the Emirate of Sharjah established the Arab Science and Technology Foundation (ASTF) in Sharjah with a $6 million donation. ASTF have the mission of identifying and supporting scientific research activities in the Arab World. A foundation dedicated solely to Emirates, the Emirates Foundation (EF), was also established by the Emirate of Abu Dhabi in 2005. EF supports five core programs: education, S&T, environment, arts & culture, and social development. In the following year, the Emirates Institute for Advanced Science and Technology (EIAST) was established by the Emirate of Dubai. EIAST is aimed at inspiring scientific innovation and fostering technological advancement. EIAST is currently tasked to initiate a space program and develop renewable and sustainable energy and water sources for Dubai. In 2007, another foundation was established by the ruler of Dubai, the Mohammed bin Rashid Al Maktoum Foundation (MBRF), with $10 billion endowment toward the development of a knowledge-based society. A foundation specifically dedicated to R&D, the National Research Foundation (NRF), was established in 2008 with a $30 million annual budget. The vision of NRF is to support research activities, and create competitive research environment and innovation system in the Emirates. NRF is tasked with the introduction of research excellence centers, strategic initiatives, and competitive awards and grants.

The United Arab Emirates University (UAEU) is the first national university in Emirates. The number of colleges grew to 9 from the original 4 when it was established in 1976. As of 2011, UAEU have over 650 faculty members, offers both undergraduate and graduate degrees to over 12,000 students. It was announced in 2009 that the Emirates Centre for Nanosciences and Nanoengineering will be established within the College of Engineering at UAEU by a $10 million grant from NRF. At the time of the announcement, there were 18 scientists working on nanotechnology research projects but the center's director stated that another 10 researchers will be recruited for the new center. The research at this center will be aimed toward cancer treatment, solar energy, and building materials.

The Higher Colleges of Technology (HCT), the largest higher educational institution in Emirates, was founded in 1988. With 13 colleges and 17 campuses throughout the country, HCT offers more than 90 programs at different levels in applied communications, business, engineering, information technology, health sciences and education. English is the official language of instruction and faculty members recruited from around the world. In 2006, the Center of Excellence for Applied Research and Training (CERT) began as the commercial arm of HCT. The $35 million investment made CERT grow to be the largest investor in the discovery and commercialization of technology in the Middle East. According to its website, CERT is the only supercomputing center in the Middle East region. CERT's Blue Gene supercomputer offers 5.7 teraflops calculating speed for use in biotechnology, nanotechnology, and genetics research as well as oil and gas simulation.

EIAST has also included a nanotechnology research laboratory within its new division Emirates Science Research Center (ESRC) for the development of renewable and sustainable energy and water sources for Dubai.

Masdar Institute of Science and Technology (MIST) is the world's first graduate-level-only university. MIST was established in 2007 with a goal to become a world-class research-driven university, focusing on advanced energy and sustainable technologies. MIST has strong ties with the Massachusetts Institute of Technology (MIT) which has supported its development and aim to become a world-class institution. MIST is the first part of the wider Masdar City Master plan to be realized a prototypical and sustainable city, one in which residents and commuters can enjoy the highest quality of life with the lowest environmental footprint. It should be noted that the $16 billion Masdar enterprise is a wholly-owned subsidiary of the Mubadala Development Company.

MIST commenced teaching in 2009 with 92 students from 22 countries and is planning to reach student population of about 800. Accepted students are offered a full tuition scholarship, monthly stipend, travel reimbursement, personal laptop, textbooks, and accommodation. Currently, MIST has 9 faculty members working in nanotechnology areas with diverse expertise. Supported by 20 students and facilities that include clean room (under construction), these faculty members work on projects related to: photovoltaic devices, biodegradable nanocomposite materials, nanoparticles, nanofluids, nanoscale transport in thermoelectric materials, nanostructured materials and their applications in emerging technologies, microfabrication and nanofabrication, low power high-performance nanoelectronics, nanophotonics, and nanomemory technologies.

Ras Al Khaimah Center for Advanced Materials (RAK-CAM) was established in 2007 by the Ruler of the Emirate of Ras Al Khaimah. RAK-CAM aims to position Ras Al Khaimah as a key contributor to the long-term technological development of the Emirates as a leader in advanced materials research. RAK-CAM research areas include: nanomaterials for diverse applications, inorganic and hybrid materials, materials for environmental remediation and hydrocarbon processing (catalysis and separations), materials for water purification and conservation, materials for solar energy applications and energy storage systems, advanced structural materials, ceramics and composites, polymeric materials and polymer nanocomposites, and biomaterials and biofuel technologies. Each research area is supported by 4-5 permanent scientists along with technical and administrative staff as well as short-term post-doctoral researchers.

RAK-CAM's research facilities include integrated state-of-the-art systems for materials synthesis and preparation, analysis, testing and characterization, together with an advanced research computing capability. The materials characterization facility include High Resolution Transmission Electron Microscope (HRTEM), Scanning Electron microscopes (SEM), Atomic Force Microscope (AFM), X-ray Powder Diffraction (XRD), X-ray Single Crystal Diffractometer, Nuclear Magnetic Resonance (NMR) spectrometers, mass spectrometry, optical spectroscopic techniques, surface characterization techniques, catalysis, and polymer characterization techniques. The materials characterization facility serves both the local and regional industry and will seek industrial collaborations in joint research projects.

Another Abu Dhabi Government initiative was the establishment of the Khalifa University of Science, Technology and Research (KUSTAR) in 2007. KUSTAR is currently running from a satellite campus in Sharjah and an interim campus in Abu Dhabi. The satellite campus was a result of a merger between KUSTAR and the 18 years old Sharjah based Etisalat University College. KUSTAR main campus is currently under construction in Abu Dhabi.

KUSTAR currently offers undergraduate and graduate degrees in engineering disciplines including: aerospace, biomedical, communication, computer, electronics, mechanical, and software. Other non-engineering disciplines such as logistics management, health sciences, homeland security and sciences will be added in the near future. KUSTAR has more than 600 undergraduate and about 40 graduate students and supported by more than 200 faculty and staff members. The university also has academic ties with Georgia Institute of Technology, Korea Advanced Institute of Science and Technology, University of Bristol, and Zayed University.

KUSTAR recently announced it will set up a nanotechnology research center. The aim of the center is to play a leading role in the establishment of nanotechnology research, development, and industry in Abu Dhabi and the UAE. The center will be dedicated to research on theoretical and experimental nanotechnology with strong emphasis on the performance attributes and functional demonstration of nanomaterials/systems, by assembling/integrating sophisticated materials, composite materials, and structures that translate to devices and systems at the macroscale. The research center is also intended to develop materials/solutions for applications in power/optoelectronic, aerospace, and diagnostic monitoring.

Kuwait

Interest in S&T started in Kuwait soon after it gained its independence from the British Empire in 1961 as demonstrated by the establishment of Kuwait University (KU) in 1966, Kuwait Institute for Economic and Social Planning in the Middle East (now Arab Planning Institute) also in 1966, Kuwait Institute for Scientific Research (KISR) in 1967, Kuwait Science Club in 1974, Kuwait Foundation for the Advancement of Sciences (KFAS) in 1976, Kuwait Institute for Medical Specialization in 1984, Ministry of Higher Education 1988, Center for Research and Studies on Kuwait in 1992, the Scientific Center and Kuwait Inventors Bureau in 2000, Dasman Centre for Research and Treatment of Diabetes in 2006, annual Kuwait Science Fair since 2008, and Sabah Al-Ahmad's Center for Giftedness & Creativity in 2009. By late 70s Kuwait had a renewable energy (solar and wind) research program and even initiated a nuclear energy program for seawater desalination and electricity production. Although lacking a formally enacted national science policy, such development made Kuwait a leading hub for S&T in the region in the 60s, 70s and early 80s. However, during mid-80s, the country suffered from terrorism (bombing of public areas, hijack of Kuwait Airways airplane, and attempt to assassinate head of state (the Emir of Kuwait). The act terrorism reached its climax in 1990 when Kuwait was invaded by its northern neighbor, Republic of Iraq. At the end of the seven month-long Iraqi occupation, around 773 Kuwaiti oil wells were set ablaze by the Iraqi army resulting in a major environmental and economic catastrophe. Kuwait's infrastructure was severely damaged during the Gulf War. Moreover, Kuwait has been recently rocked by a series of political crises. In the last five years, seven governments have resigned and the parliament has been dissolved three times. This took a toll on the country's development and delayed many vital projects.

At the time of its establishment, KU had four colleges, namely College of Science, College of Arts, College of Education, and College for Women. At that time, KU had 418 students enrolled and 31 faculty members. In 1967 another five colleges were added: College of Law, College of Islamic Studies, College of Business, College of Economics, and College of Political Science. College of Medicine, College of Engineering and Petroleum, and College of Graduate Studies were added in 1973, 1974, and 1977 respectively. College of Allied Health Sciences and Nursing, College of Pharmacy was added in 1982 followed by College of Pharmacy and College of Dentistry both in 1996. By 2005, the KU grew to more than 19,000 students and over 1,000 faculty members.

When KISR was first established, it was tasked to carry out applied scientific research in three fields: petroleum, desert agriculture and marine biology. KISR is directly responsible to the Council of Ministers. In 1981 KISR formally became independent public institution governed by the Board of Trustees chaired by the minister of Education. KISR's objectives are to carry out applied scientific research that helps the advancement of national industry and to undertake studies relating to the preservation of the environment, resources of natural wealth and their discovery, sources of water and energy, methods of agricultural exploitation and promotion of water wealth. KISR was entrusted with undertaking research and scientific and technological consultations for both governmental and private institutions in Kuwait, The Gulf region and the Arab World.

KISR took the initiative of preparing a draft for national science policy, including studying the national base, and the needs of different sectors in the field of S&T in Kuwait. The document entitled "National Policy for Science, Technology and Innovation of the State of Kuwait" was reviewed by experts from Kuwait, Arab and foreign countries including the United Nations Economic and Social Commission for Western Asia (UN-ESCWA) and finalized for submission to the competent authorities in Kuwait in 2007[11]. This made Kuwait one step ahead of most Arab States toward national science policy framework. The national policy draft acknowledged the importance of nanotechnology for Kuwait's S&T development.

KFAS establishment was initiated by the former Emir of Kuwait when he was Crown Prince in an attempt to encourage and support scientific research. KFAS is managed and administered by a Board of Directors, chaired by the Emir of Kuwait. The Board is comprised of six members appointed by the Emir for a period of three years. Article (6) of the Memorandum of Association of KFAS states that a source of KFAS's funding shall be paid by all Kuwait Shareholding Companies (KSC). Each KSC Company shall pay 5% of its net profits to KFAS. The compulsory contribution to KFAS from KSC has been reduced to 2% in 1999 and now is at 1%. KFAS's goal is to promote scientific, technological and intellectual progress within Kuwait and the region by providing financial sustenance to research in basic and applied sciences, supporting projects of national priority, awarding prizes and recognition at national, regional and international levels, organizing scientific symposia and conferences, enriching the Arabic language library by publishing journals, books and encyclopedias, and promoting scientific and cultural awareness.

In 2008 the government introduced the new five-year development plan (2009-2014). The five year plan is the first in a series of five such plans, stretching to 2035, which aims to convert Kuwait to a trade and financial hub of the region. Kuwait also contacted a private consultancy firm, Tony Blair Associates, to develop a vision for Kuwait development up to 2035. In 2009 a report entitled "Vision Kuwait 2030" was delivered by UK former prime minister to Kuwait authorities.

Kuwait Development plan (KDP) is considered a first of a kind plan since 1986. Following years of sub-par government spending on infrastructure and development projects, the National Assembly by a unanimous vote approved the spending of $129 billion on KDP in 2010[12]. About $6 billion has been allocated for the development of Sabah Al-Salem University City, an educational complex spreading over a 6 km² area. It will consist of three campuses including 16 colleges and support faculties, a hospital, a hotel, a housing complex, sports facilities, and auditoriums. Upon its completion in 2020, the city will have the capacity to accommodate up to 40,000 students.

Also in 2010 a $250 million initiative was launched by the Emir of Kuwait to support R&D projects in renewable energy, peaceful uses of nuclear energy, food science, water resources, etc. KISR was able to secure about $50 million of the initiative budget to establish nanotechnology center. KISR's interest in nanotechnology dates back to 2006 when it held Kuwait's first nanotechnology conference. KISR Nanotechnology Research Center (KNRC) research focus areas include renewable energy systems (photovoltaic, fuel cell, and hydrogen storage), construction materials (high performance concrete), surface protection coating materials (corrosion and erosion resistant, self-cleaning, and antibacterial), catalyst materials (oil production and refining), water purification and desalination, and chemical and physical sensing technologies. KNRC will house state-of-the-art 360 m² clean room facility equipped fabrication and characterization tools, materials synthesis laboratory, modeling and simulation laboratory, and chemical and physical properties characterization facilities.

Back in 1976 KU College of Science established an Electron Microscopy Unit (EMU) which evolved over the years to be known now as Nanoscopy Science Center (NSC). The center offers many services biological and material science research such as transmission electron microscopy (TEM), HRTEM, SEM with energy dispersive spectroscopy (EDS) capabilities, laser confocal scanning microscopy, AFM, freeze-fracture machine, sample preparation ultra microtomy, glass knife makers, tissue processors, cryo-station, sputter coater, and critical point dryer NSC also has special labs for biological and material sciences preparations and two fully equipped interconnected darkrooms. Recently, NSC acquired a variable pressure remotely operational field emission SEM fitted with EDS and high resolution cryo-transfer system. NSC is managed by five faculty members and seven fulltime staff members. The college of engineering and petroleum at KU established in 2007 the Kuwait University Nanotechnology Research Facility (KUNRF) to service nanotechnology R&D in the college. KUNRF is managed by four faculty members, employs four technicians, and three fulltime professional research assistants. The facility houses a clean room with several fabrication and characterization tools. Most of KU nanotechnology research fall in the fields of photovoltaic, nano-electronics, biotechnology, and advanced materials.

Qatar

According to its national development strategy 2011-2016 document[13], Qatar has recently invested considerable resources in R&D. An outstanding infrastructure is in place for scientific research, with programs to draw potential researchers and build partnerships with universities and businesses. However, similar to most Arab States, Qatar does not have formally enacted national science policy. The country's R&D took a sharp turn when Qatar's current head of state (the Emir of Qatar) assumed power in 1995. In the same year, he established Qatar Foundation for Education, Science and Community Development (QF). The aim of this foundation is to unlock the human potentials through its three pillars of education, science & research, and community development. Since its inception, Qatar's First Lady acted as QF's chairperson and has been its driving force. In 2006, the Emir of Qatar pledged to allocate 2.8% of Qatar's GDP to science and research which translates to about $1.5 billion per year. QF is tasked to manage this budget. Moreover, in the same year, he issued a decree to establish the general secretariat for development planning to coordinate plans, strategies and policies in support of Qatar's National Vision 2030 (QNV2030). Approved in 2008, Qatar's long-term development strategy defines broad future trends, sets goals, and provides the framework for Qatar's National Development Strategy. QNV 2030 rests on four pillars: human development, social development, economic development, and environmental development. However, there is no explicit mention of nanotechnology in QNV 2030 or in Qatar National Development Strategy 2011-2016.

According to Schwab[14], Qatar has topped other Middle Eastern countries as it was ranked 14th on Global Competitiveness Report (GCR). Topping 139 countries included in the GCR, Qatar move up from the 22nd rank in 2009 to the 14th rank for 2011. As stated by the report, competitiveness in Qatar is based on strong fixed pillars demonstrated in highly-efficient and well-performing institutional platforms, stable economic environment and effective commodities market. Furthermore, Qatar's human development index moved from 55 in 1997[15] to 38 in 2010[16] out of 169 countries according to the Human Development index published by the United Nations Development Program.

Qatar has invested heavily in developing capabilities for scientific innovation and research. QF has established a broad range of research centers within Qatar University and Education City, including research opportunities in scientific and technical areas; policy, social and business areas; innovative design and culture and heritage. QF also paid attention to education programs on research at the university as well as K–12 level. In addition to graduate level research programs, undergraduate research experience program and specialized math and science tracks in secondary schools were established.

Qatar National Research Fund was established in 2006 to accelerate quality R&D by providing research grants to a wide range of beneficiaries. Qatar Science and Technology Park (QSTP) was established in 2009 to attract investments from several international businesses for frontier R&D. European Aeronautic Defense and Space Company (EADS), ExxonMobil, General Electric (GE), Microsoft, Shell, Total and others have already committed $225 million of R&D investment at QSTP.

Education City is QF's flagship project. Located on the outskirts of Doha, the capital of Qatar, the city covers 14 km^2 and houses educational and research facilities from primary school to graduate level. Education City aims to be the center of educational excellence in the region. It is conceived of as a forum where universities share research and forge relationships with businesses and institutions in public and private sectors. Education City is home to branch campuses of six international universities which include Weill Cornell Medical College, Virginia Commonwealth University School of the Arts, Carnegie Mellon University offering programs in business administration, biological sciences, computational biology, computer science and information systems, Texas A&M's School of Engineering, Georgetown University School of Foreign Service, and Northwestern University School of Communication and Medill School of Journalism.

QF also planning to establish several applied research centers of excellence in Qatar including: Center for Genomic and Proteomics Medicine, Center for Stem Cells Research, Center for Molecular Imaging Research, Center for Infectious Diseases, Center for Bioinformatics and Data Mining, Center for Applied Nanotechnology, and Center for Environmental Research.

Current nanotechnology R&D activities in Qatar are faculty members driven with some research focused on: catalyst for natural gas liquefaction, nanoparticles for cancer treatment, and nanomaterials synthesis including functional nanofibers for protective textile applications, water filtration, and biomedical applications.

Saudi

Saudi leaders have recognized the importance of harnessing S&T for their developmental needs as early as 1977 when they established the Saudi Arabian National Center for Science and Technology (SANCST). The center was tasked to conduct applied research, manage scientific manpower, develop a national science policy, award scholarships and grants, and coordinate between the different scientific institutions and the central government. In 1985 SANCST was renamed as King Abdul Aziz City for Science and Technology (KACST) to become Saudi's principal agency for promoting scientific and technological R&D. KACST was directed by its charter of 1986 to propose a national policy for the development of S&T. However, preparations for the national policy for science and technology did not start until mid-1997. KACST in cooperation with the Ministry of Economy and Planning (MOEP) developed a long-term national policy for S&T. In July 2002 the Council of Ministers approved the National Policy for Science and Technology (NPST) under the name of "The Comprehensive Long-Term National Policy for Science and Technology" making Saudi one of the few Arab States to have such policy. Saudi's NPST included a timetable for gradually increasing sources of funding R&D, which is to reach 1.6% of the GDP in 2020. KACST was also put in charge of supervising the implementation of the policy.

KACST, MOEP, and other relevant stakeholders, developed the national plan for STI under the framework of the NPST. The plan outlined the focus and future direction of STI in Saudi, with special consideration of the role of KACST, universities, government, industry and the society at large. The plan encompasses eight major programs: strategic and advanced technologies, scientific research and technical development capabilities, transfer, development and localizing technology, science, technology and society, scientific and technical human resources, diversifying financial support resources, STI system, and institutional structures for STI. Within the strategic and advanced technologies program, KACST is responsible for five years strategic and implementation plans for 11 technologies: water, oil & gas, petrochemicals, nanotechnology, biotechnology, information technology, electronics, communication, & photonics, space and aeronautics, energy, environment, and advanced materials. Each plan establishes a mission and vision, identifies stakeholders and users, and determines the highest priority technical areas for Saudi.

The mission of the National Nanotechnology Program (NNP) is to ensure that Saudi is a major player within the international community in R&D of nanotechnologies. NNP will foster academic excellence and ensure that world-class R&D facilities are available to all parts of the economy, from academic institutions to industry. NNP is envisioned to create a multidisciplinary program leveraging all branches of science in order to build competence and capability in nanotechnologies that will help to ensure the future competitiveness of Saudi[17].

To foster the growth of nanotechnology, the NNP will strengthen academic research, improve infrastructure, link research with economic and industrial strategy, create an international collaboration plan, create a management plan, developing health, safety, and standards/processes plans, strengthen education and workforce plans, and developing a commercialization plan. The NNP has identified three nanotechnology subtopics: quantum structure and nanodevices, nanomaterials and synthesis, and computational modeling and theoretical analysis of nanosystems. There are several areas that are strategically important to Saudi and are expected to benefit from NNP such as: improved desalination, enhanced catalysis, corrosion resistance, sensing nanodevices, solar cells, enhanced oil recovery and well productivity, medical diagnosis and drug delivery, electronic, and photonic nanodevices, and MEMS.

Currently, much of the expertise and many of the facilities for conducting nanotechnology research are located at KACST and the following universities: King Fahd University of Petroleum and Minerals (KFUPM) established in 1963, King Abdul Aziz University (KAU) established in 1967, Riyadh University established in 1957 but was renamed to King Saud University (KSU) in 1982, King Abdullah University of Science and Technology (KAUST) established in 2009 with $10 billion endowment (6th largest in the world), King Khalid University (KKU) established in 1998, King Faisal University (KFU) established in 1975, and Taibah University (TaibahU) established in 2003. It is estimated that approximately 30 research projects in the field of nanotechnology have been launched at the above universities and research institutes. Industry is also take advantage of nanotechnology research. Local companies such as Saudi national oil company (established as Arabian American Oil Company, known now as Saudi ARAMCO) and Saudi Basic Industries Corporation have devoted resources to conducting nanotechnology research. The two companies alone have launched more than 20 research projects in the field of nanotechnology. To support this research, they have employed more than 20 PhDs on staff with expertise applicable to nanotechnology research. Much of this research has been in materials and synthesis. While the application of this research has often been aligned with the industrial and economic needs, for instance, looking at improving fossil fuel extraction with nanomaterials, some research has looked at other nanotechnology applications such as: structural materials and coatings, biotechnology, catalysis and membranes, sensors and measurement, electronics and magnetics, energy and environment.

KACST has already formed R&D partnerships and collaborative programs with some leading international institutions such as KACST/IBM Nanotechnology Center of Excellence to cover research in water desalination, catalysts for petrochemical applications, and solar energy, University of Auckland, focusing on the development of nano light emitting diodes (LEDs), MIT/KACST/KFU projects which include: nanopatterning of fuel cells electrodes, enhancement of transport phenomenon using nanofluid, photoacoustic detection system for the petrochemical industry, mid-infrared laser for sensing applications, University of Minnesota/KACST for formation of titanium oxide nanotube, University of Illinois, Urban-Champain/KACST for development of electrochemical cell used for silicon nanoparticles formation, University of Michigan, Ann Arbor/KACST for using nanoimprint to develop inexpensive solar cells, National Academy of Sciences of Belarus for carbon nanotube production facilities and development, production and installation of a scanning probe microscope.

Saudi universities have been steadily increasing the number of their international partnerships with the hope of broadening their research and expertise. These collaborations include: In 2005, KFUPM sent three faculty members to National University of Singapore to explore and initiate research collaborations. In 2007, KFUPM and MIT announced preparations to inaugurate a scientific collaboration agreement in the field of education and scientific research between the Mechanical Engineering Departments of both institutions. KAUST announced a partnership with the Indian Institute of Technology, Bombay (IIT Bombay). This partnership will involve collaborative research in many areas related to nanotechnology. KAUST and AUC agreed to collaborate in many research areas including nanotechnology and advanced materials.

Saudi ARAMCO is funding and collaborating with the Australian Research Council's Centre of Excellence for Functional Nanomaterials on a four-year research project to develop catalytic materials. Saudi ARAMCO also has a contract with a Canadian company for planning, implementing, and carrying out a product development program entitled "Application of Nanotechnology for In-Situ Structural Repair of Degraded Heat Exchangers."

Saudi has established several nanotechnology research centers even before the full realization of the NNP. The Center of Excellence in Nanotechnology (CENT) at KFUPM was established in 2005. CENT currently employs five faculty members, one research scientist, three post-doctoral fellows, one lecturer, two engineers, two scientists, and two administrative staff. CENT also has twenty two affiliated faculty members from physics, chemistry, and engineering departments within KFUPM. CENT's main research focus is on catalysis and photo-catalysis, nanostructured chemical sensors, and carbon nanotubes production and applications. It also conducts activities in the field of anticorrosion processes, biotechnology, environment, and solar cells. CENT is equipped with various state-of-the-art instruments ranging from compositional analyses and physical properties measurements to sintering and synthesis of wide ranges of materials, such as; metals, ceramics, intermetallic, composites etc. Some of the equipment listed on CENT website include cold isostatic presses, hot isostatic press, field assisted sintering technique, hydraulic press, brabender mixer, thermal analyzer (TG-DSC), nano/micro particles analyzer, microwave oven for sintering, automatic sputter coater, autoclave minireactor, tubular furnace, ultra liquid processor, gas chromatography system, microwave synthesis lab station, and pulsed laser deposition. CENT scientific production started in 2007. The center holds 2 patents with 4 pending and published more than 60 articles in international journals, two book chapters, and presented more than 37 papers at various international conferences.

The Center of Nanotechnology (CNT) at KAU was established in 2006. The multidisciplinary center work covers several science and engineering areas such as: engineering, pharmacology, medical sciences, genetic engineering (Artificial DNA), basic sciences e.g. physics, chemistry and biology, material science, MEMS devices, computational nanotechnology, fabrication & assembly of different nanomaterials, advanced materials including polymers and semiconductor nanomaterials, and safety and health effect of nanoparticles. CNT carry out its research activities through ten research groups. Each group consists of affiliated researchers from different departments and collages at KAU. The groups name and their size are: nanomaterial (36 researchers), nanofabrication (30 researchers), nanocomposites (23 researchers), nanobiotechnology (55 researchers), nano drug delivery (49 researchers), nanomedicine (61 researchers), nanotechnology-based renewable energy (28 researchers), nanodevices & nanosystems (26 researchers), nanotechnology for desalination & water treatment (30 researchers), nano-computation and simulation (17 researchers). CNT supports its research with several facilities which include: nanomeasurements laboratory, nanomaterials synthesis laboratory, nanofabrication laboratory, and microscopy laboratory. As for scientific production, CNT first journal article appeared in 2007 and so far, has accumulated more than 32 articles published in international journals.

In 2009, the Center of Excellence of Nano-manufacuturing Applications (CENA) was established at KACST. CENA is a research consortium between KACST, Intel, and selected universities in the Middle East and North Africa (MENA) region. The objectives of the consortium are to build up regional human capacity and reverse the brain drain in the MENA region. The universities include Saudi universities (KFU, TaibahU, KFUPM, and KSU), Egyptian universities (Alexandria University, Arab Academy for Science and Technology and Maritime, and Assuit University), Lebanese university (American University of Beirut), Palestinian university (AnNajah National University), and Turkish universities (Middle East Technical University and Istanbul Technical University). Other universities/research institutions from Algeria (three universities), Egypt, Jordan (two universities), Morocco, Tunisia and Emirates (three universities) are in the process of joining CENA. Research in CENA is focused on three main themes: fabrication of MEMS, sensors, and integration of CMOS-MEMS, fabrication of autonomous RF sensors and applications, and nano-materials synthesis, processing, and characterization. CENA is planning to construct 3500 m^2 center with state-of-the-art equipment to enable research in nanoprocessing, III-V materials and devices, and optoelectronics. The facility will have lithography, epitaxy, thin film deposition and thermal processing, dry etching, metrology, wet processing, and backend capabilities.

One of KAUST's core facilities is the Advanced Nanofabrication, Imaging and Characterization (ANIC) facility. ANIC is dedicated to providing the instrumentation, technical expertise, and team-teaching to stimulate collaborative research in nanoscale technology. ANIC is a complex of multidisciplinary laboratories that supports research across many different departments within KAUST. ANIC supports not only materials and device research in physics, electrical engineering, mechanical engineering and chemistry, but it also facilitates research interaction and collaboration between the physical, chemical, biological and medical disciplines. ANIC laboratories are grouped into two sub-facilities, advanced nanofabrication sub-facility and imaging and characterization sub-facility. The nanofabrication sub-facility occupies 2000 m^2 of Class-1000 clean room space with multiple bays at Class-100. It includes capabilities for device fabrication and characterizations with a range of equipment, such as plasma etching, chemical vapor deposition, metallization, wet chemical stations, thermal (oxidation/diffusion) and lithography. The sub-facility is divided into 5 modules, namely: lithography and mask making, deposition and thermal diffusion, wet and dry etching, metrology, and package. This sub-facility is managed by 4 research scientists, 3 research engineers, and 7 technicians. The imaging and characterization sub-facility has comprehensive capabilities for scanning, transmission, confocal, and Raman microscopy, magnetic and thermal measurements, and other instrumentation for materials characterization. It also houses a NMR laboratory which comprises of a suite of 10 NMR spectrometers for solution-based and solid-state samples, together with comprehensive sample preparation facilities for the study of macromolecular structures and spatial distributions, dynamics in solution, and chemical composition of small features in solid-state samples. The sub-facility is divided into 9 modules, namely TEM, SEM, optical microscopy, surface science, XRD and X-ray fluorescence, thermal analysis, low temperature physics laboratory, NMR spectroscopy, and microwave testing laboratory. This sub-facility is managed by 10 research scientists, 2 research engineers, and 7 technicians.

Most recently, King Abdullah Institute for Nanotechnology (KAIN) was established in 2010 at KSU. KAIN's conducts a wide range of activities spanning areas such as research, development, and applied activities (energy, water treatment and desalination, telecommunications, medicine and pharmaceutical, food and environment, and manufacturing and nanomaterials), modeling and simulation of nanomaterials, fields of education and training, and economic and industrial fields. To support these research activities, KAIN is planning to employ 15 group leaders, 30 researchers, 30 assistant researchers, 15 technicians, and 60 graduate students. A 13,000 m^2 has been designated to KAIN in which 8,000 m^2 will be used for building laboratories, administrative and researcher's offices, warehouses, workstations, and service areas. A budget of $20 million has been allocated to establish a clean room equipped with state-of-the-art equipment for nanotechnology research activities. KAIN researchers are currently working on 23 different projects and, to date, 9 patents have been files in Europe and U.S. patent offices and about 20 papers have been published. KAIN also established links with other Saudi universities pursuing nanotechnology as well as international institutions such as University of Illinois, Ohio State University, Malaysian University of Science, Lena Nanoceutic, Innovatecs, Quadrtk Center, Illinois Institute of Technology, Max Plank Institute, Chinese Academy of Science, University of Virginia, and Virginia Commonwealth University.

CONCLUSION

Arab States, though share common language, similar cultures, and even R&D priorities, they vary in terms of wealth, political structure, and S&T development stage. It's clear that governance plays an important role in the advancement of S&T. In our five countries sample, two have parliamentary system (Egypt and Kuwait) and three have absolute monarchy. Generally speaking, S&T infrastructure development started earlier in Egypt and Kuwait. Saudi, Emirates, and Qatar had to wait for the executive decision but experienced exceptionally rapid growth that surpassed Egypt and Kuwait in many areas especially nanotechnology infrastructure development. With NPST and NNP in place, Saudi quickly achieved advanced nanotechnology R&D stage relative to others. Egypt's rich abundance of human resources allowed it to lead the scientific production in nanotechnology even with its limited infrastructure and financial resources. Emirates and Qatar are rising stars in nanotechnology and anticipated to produce significant R&D in nanotechnology once they build their human capacity. There are tremendous opportunities in nanotechnology R&D for students, researchers, technology providers such materials suppliers and equipment manufacturers, as well as startup companies in the five countries investigated.

ACKNOWLEDGMENT

The authors would like to thank Dr. Khaled Saoud of Virginia commonwealth University-Qatar for providing information on nanotechnology activities in Qatar. This work was funded by KISR project EA038C.

REFERENCES
[1] J. F. S. Jr., The National Nanotechnology Initiative: Overview, Reauthorization, and Appropriations Issues, Congressional Research Service RL34401, (2011).
[2] Strategy for Science, Technology and Innovation in Islamic Countries, Islamic Educational, Scientific and Cultural Organization (2009).

[3]Unesco Science Report 2010: The Current Status of Science around the World, United Nations Educational, Scientific and Cultural Organization, Paris, France (2010).

[4]W. Sawahel, Building Scientific and Technological Talent in the Broader Middle East and North Africa Region (Bmena) ...... Arab Capability Status & Present Development and Proposed Action Plan, IDB Science Development Network (2010).

[5]W. A. Sawahel, Higher Education and Science & Technology in Idb Member Countries Present Development and Future Prospects, IDB science development network (2008).

[6]M. Yahaya, M. M. Salleh, I. Ho-Abdullah, and Y. C. Chin, Roadmap for Achieving Excellence in Higher Education in Nanotechnology, Islamic Development Bank (2009).

[7]P. Hahn and G. M. z. Köcker, The Egyptian Innovation System: An Exploratory Study with Specific Focus on Egyptian Technology Transfer and Innovation Centres, Institute for Innovation and Technology (2008).

[8]Science and Technology Policy, Research Management and Planning in the Arab Republic of Egypt, U.S. National Academy of Sciences (1976).

[9] Science, Technology and Innovation (Sti) System in Egypt, Academy of Scientific Research & Technology (2010).

[10]Uae Vision 2021, UAE Federal Cabinet (2010).

[11]N. Al-Awadhi and Y. Al-Sultan, National Policy for Science, Technology and Innovation of the State of Kuwait, Kuwait Institute for Scientific Research (2007).

[12]Development Projects...Gateway to Future of Kuwait, Ministry of Information (2011).

[13]Qatar National Development Strategy 2011~2016, Qatar General Secretariat for Development Planning (2011).

[14]K. Schwab, The Global Competitiveness Report 2011-2012, World Economic Forum (2011).

[15]Human Development Report 1997, United Nations Development Program (1997).

[16]Human Development Report 2010, United Nations Development Program (2010).

[17]Strategic Priorities for Nanotechnology Program, King Abdulaziz City for Science and Technology and Kingdom of Saudi Arabia Ministry of Economy and Planning (2007).

FINITE ELEMENT MODELING FOR MODE REDUCTION IN BUNDLED SAPPHIRE PHOTONIC CRYSTAL FIBERS

Neal T. Pfeiffenberger and Gary R. Pickrell

Virginia Tech
Blacksburg, VA, USA

ABSTRACT

This paper presents the finite element modeling of a unique bundled sapphire photonic crystal fiber design. The structure generally consists of six rods of single crystal sapphire fiber symmetrically arranged around a solid single crystal sapphire core. The single crystal sapphire fibers used in this study were approximately 50μm and 70μm in diameter as this is the present limitation of single crystal sapphire fiber available for purchase. The modeling work focuses on the optimization and modal analysis of this photonic crystal fiber using Comsol Multiphysics 4.1. The goal is to reduce the modal volume of the fiber to limit the loss during end use applications such as pressure and temperature sensing in coal gasifiers and oil wells.

INTRODUCTION

Single crystal sapphire, also known as α-Al_2O_3, is a widely sought after material for use in optical fibers. This is due to single crystal sapphire's ability to survive high temperatures and harsh environments. Single crystal sapphire has a melting temperature of 2053°C making it very useful for high temperature applications. Its single crystal nature ensures that it has no grain boundaries, which improves the chemical resistance to corrosion. Silica fibers, which are commonly used for pressure and temperature sensors as well as for communication, are cheaper and more readily available compared to single crystal sapphire. Silica, however, is limited by a creep and plastic deformation at temperatures near 900°C[1]. Most silica fibers have a doped core and at 1000°C, dopant migration from the core becomes appreciable[1]. The dopant migration changes the confinement properties of the fiber, which limits the upper use temperature for sensing applications.

Single crystal sapphire also operates over a wide wavelength range from 0.24 - 4.0μm[2] with a corresponding refractive index range from 1.785 – 1.674. The infrared region is especially important for high power applications. Most single crystal sapphire fibers are highly multimoded due to their inability to be clad during the production stage. All cladding is applied to these fibers after fabrication. Some of the materials that have been demonstrated include SiO_xN_y, Mg_xSiO_y, Ti_xSi_yO, polycrystalline alumina, silicon carbide, zirconia, and niobium with the x and y values having the ability to be changed to alter the refractive index of the coating[3]. The ability to clad an optical fiber is important for a variety of reasons. One of the main reasons is the reduction of modal volume, which would provide a number of benefits especially in the case of interferometric sensors. Reducing the modal volume makes the production of interferometric sensors much easier and reduces the sensitivity of the sensor to other noise factors. Reducing the modal volume can also reduce the loss of the fiber (all other things being equal) and providing a cladding also reduces the loss due to contaminants depositing on the surface of the fiber. The eventual goal for these fibers would be to approach single mode operation. Other benefits of a fiber cladding include improving the mechanical stability of the fiber.

The fiber designs examined in this paper have an outer ring of single crystal sapphire fibers which surround the central core and produce voids periodically around the central core, which lowers the effective refractive index of the air-sapphire region surrounding the core. This allows for a modal reduction while still maintaining the material advantages of using single crystal sapphire as a cladding layer as previously described. The first photonic crystal fiber made of single crystal sapphire fiber was published previously by our group[4] and incorporated this same hexagonal shape in a 15cm long bundled fiber. Initial work has been accomplished with the multiphysics models to examine the modal structure of these types of fibers in greater detail[5].

The fibers examined in this study are expected to fall into the photonic crystal fiber category. These fibers all operate via total internal reflection (TIR). In these fibers, light is more strongly confined to the core due to a large refractive index difference of single crystal sapphire, 1.74618 vs. 1.0 for air at a free-space wavelength of 1.55μm. The other branch of photonic crystal fibers operate via a photonic band gap where a periodic arrangement of a material with a given refractive index difference between that of the background material will not allow electromagnetic waves of a certain range of wavelengths to be transmitted. If one of the periodic holes in the lattice is removed, this will confine a range of wavelengths from operating anywhere besides in this defect region[6] if the size and/or spacing of the holes is in correct relation to the wavelength range of the injected light.

The fibers in this study examine how the rod diameter of a hexagonal ring of single crystal sapphire affects the modal volume of the fiber. In this study two different fibers are compared versus a single rod case with no cladding. The number of modes that propagate in an optical fiber is directly affected by the size of the core. The core sizes in these cases are 50μm and 70μm in diameter, respectively. The number of modes propagating in a step index multimode fiber, Nm, can be approximated[7] for weakly guiding fibers with large a number of modes by Equation (1).

$$Nm = 0.5 \times \left(\frac{\pi + D * NA}{\lambda} \right)^2 \qquad (1)$$

Where D = diameter of the core, λ = wavelength of propagation and NA is the numerical aperture, which is the light gathering ability of the fiber. The equation for numerical aperture can be approximated by Equation (2).

$$NA = \sqrt{n_f^2 - n_c^2} \qquad (2)$$

With n_f = the refractive index of the core and n_c = the refractive index of the cladding. In the single rod case, n_f = 1.74618 (single crystal sapphire) while n_c = 1.0 (air). In both bundled fibers, n_c = an average somewhere between 1.0 and 1.74618, lowing the effective refractive index of this material. Comsol Multiphysics estimates the effective refractive index of this air-sapphire region surrounding the core in both fibers to be approximately 1.7147. Plugging this value into Equation (2) for n_c with n_f = 1.74618 we obtain an NA of 0.330. When this value is substituted in Equation (1) above we obtain Equation (3)

$$Nm = 0.5 \times \left(\frac{\pi + D * (0.330)}{1.55\mu m} \right)^2 \qquad (3)$$

When D=70μm, Nm = 1096.05 and when D=50μm, Nm = 559.209. The real question that this modeling work aims at understanding is how the average effective index really is affected by the reduction in fiber diameter for a bundled fiber from 70μm to 50μm at 1.55μm.

This paper examines two single crystal sapphire photonic crystal fibers through the use of a Finite Element Modeling (FEM) program called Comsol Multiphysics 4.2. FEM is useful for fiber optics research because it can accurately predict the modal structure of an optical fiber without fabrication. The objective of this paper was to model the propagation characteristics of both a bundled fiber with 70μm diameter rods and a bundled fiber with 50μm diameter rods at 1.55μm with the rods all being composed of single crystal sapphire. This process allows us to examine the benefits of a smaller core size for this bundled structure. This paper is an extension of the previously published work by our group[5] which examines alternate fiber designs for later fabrication.

FEM ANALYSIS

The fibers in the following sections are composed of a single crystal sapphire rod surrounded by a ring of 6 other single crystal sapphire rods, all of which are either 70μm (left) or 50μm (right) in diameter as seen in Figure 1. below. The air regions, as seen in blue in Figure 1, have a refractive index of n=1.0 while the gray region has a refractive index of n=1.74618. The process for creating this type of fiber experimentally has been outlined in previous papers[4,5].

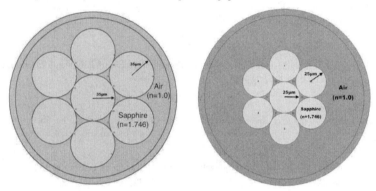

Figure 1. Schematic of fibers used in FEM. At left, the bundle composed of 7 (70μm) rods of single crystal sapphire (in gray) surrounded by air (blue region). At right, the bundle composed of 7 (50μm) rods of single crystal sapphire (in gray) surrounded by air (blue region).

The sapphire rods in this study were examined along the C-axis of the crystal extending along the length of the fiber (z-direction). The hexagonal crystal structure of single crystal sapphire has two axes perpendicular to the c-axis, referred to as the a-axes.

Finite element analysis can be a very useful tool for fiber optic design. The program used in this paper, Comsol Multiphysics, uses high-order vectorial elements composed of both an automatic and iterative grid refinement calculator. This allows for optimum error estimation during the solving process. Both models in this paper require a complex mesh to render the large air-sapphire regions with high accuracy. These types of RF modal propagation problems are generally solved with a linear solver so that Maxwell's equations in the FEM discretization can be computed without error. Comsol Multiphysics 4.2 also features the Multifrontal Massively Parallel sparse direct Solver (MUMPS) solver, which allows for parallel computing on a quad core computer to decrease computation time.

The first step in the modeling process is the materials selection. As described above in Figure 1, two fibers, a bundled fiber with 70μm diameter rods and a bundled fiber with 50μm diameter rods are to be examined. The dimensions of the rods and the material properties are then adjusted so that the rods of single crystal sapphire (α-Al$_2$O$_3$ in gray) have a refractive index of n = 1.74618 and are surrounded by air (blue region), where n=1.0. All models in this paper are solved for using a free space wavelength of 1.55μm, as this is a standard telecommunications wavelength. The outer region of the fibers in Figure 1 is set as a perfectly matched layer (PML). The outer diameter for the PML layer is set to 280μm with an inner PML diameter of 260μm. This PML is defined as a reflectionless outer layer, which is used in FEM to absorb all outgoing waves. The PML absorbs all outgoing radiation[8] and is reflectionless for any frequency. Since no modes of interest are expected in this outer air layer, the PML region also saves valuable memory space and decreases computation time.

Confinement loss, L_c, is a commonly used term for measuring loss under optimal operating conditions in FEM modeling. It is directly related to the imaginary portion of the propagation constant[9], γ through the equation:

$$L_c = 8.686 * \alpha \qquad (4)$$

where α is the attenuation constant. This equation gives the loss[10] in units of dB/m. The propagation constant is defined as:

$$\gamma = \alpha + j\beta \qquad (5)$$

where β is the phase constant. Another important metric that Comsol Multiphysics can calculate for a given mode is the power flow time average, Po_{iav}.

$$Po_{iav} = \frac{1}{2} * Re(\mathbf{E} \times \mathbf{H}^*)_i \qquad (6)$$

which gives us the z_i component in the direction of propagation (into the fiber). This shows that at any point in an electromagnetic field, the vector in Equation (6) can be interpreted in terms of power flow in a specific direction.

The boundary conditions governing each fiber rod in the bundle is then determined. A perfect electrical conductor (PEC) setting was selected for the inner ring surrounding the fiber bundle. A PEC is a surface where the tangential component of the electric field vector ($\mathbf{E}_t$) is removed. The regions inside of the PEC where set to a continuity boundary condition. As discussed previously, a cylindrical PML was set for the region surrounding the PEC. Comsol is then solved using a direct linear equation solver (UMFPACK) in combination with the MUMPS solver, which allows for the use of multiple cores, near the effective mode index of the single crystal sapphire (1.74602 at 1.55μm).

The next step in the Comsol Multiphysics modeling process is the mesh refinement. The accuracy of the output data is a direct correlation to how precise the mesh is for each fiber structure. The mesh is limited by the memory of the computer as the degrees of freedom increase with a decrease in mesh size. The models for this paper were computed on a 12Gb Macintosh I7 running Comsol 4.2 with a solution time near 7200 seconds for 500 modes. Figure 2 below shows a mesh consisting of 64240 elements for the bundled fiber with 70μm diameter rods and 88864 elements for the bundled fiber with 50μm diameter rods.

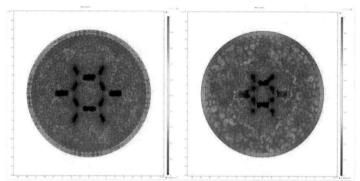

Figure 2. 2-D Comsol meshes of both the bundled fiber with 70μm diameter rods (left) and the bundled fiber with 50μm diameter rods (right).

Figure 3 below is the resultant fundamental-like hybrid mode for the bundled fiber with 70μm diameter rods solved at a free space wavelength of 1.55μm with 416663 degrees of freedom. Figure 4 below is the resultant fundamental-like hybrid mode for the bundled fiber with 50μm diameter rods solved at a free space wavelength of 1.55μm with 622176 degrees of freedom. In both Figures 3 and 4, the electric field polarization can be seen by the red arrows. The polarizations of these modes are 90 degrees from each other for a given LP mode. The optical power is expected to be concentrated directly in the center of the core of central rod in both fibers due to the coupling from the outer ring of rods in both cases. The 70μm diameter rods case has an effective mode index of 1.74602 with a corresponding confinement loss L_c= 2.0166e-8 dB/km. The 50μm diameter rods case has an effective mode index of 1.74602 with a larger corresponding confinement loss L_c= 1.92e-9 dB/km.

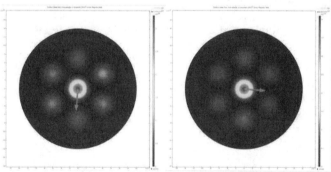

Figure 3. Fundamental LP_{01}–like modes for the bundled fiber with 70μm diameter rods showing both electric field polarizations with red arrows.

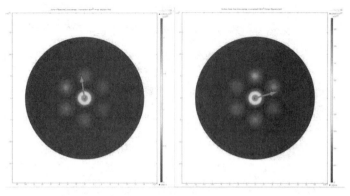

Figure 4. Fundamental LP_{01}–like modes for the bundled fiber with 50μm diameter rods showing both electric field polarizations with red arrows.

Comsol Multiphysics also allows us to calculate the number of modes that will propagate in both fibers. By solving for all of the eigenmodes at a free space wavelength of 1.55μm we find that the bundled fiber with 70μm diameter rods holds approximately 38000 modes with 5429 of these modes being confined to the central 70μm diameter core rod. The bundled fiber with 50μm diameter rods holds approximately 22058 modes with 2844 of these modes being confined to the central 50μm diameter core rod. This gives us a modal reduction of 62.49% from the bundled fiber with 70μm diameter rods to the bundled fiber with 50μm diameter rods, which is on the scale of that mentioned in Equation (3). Figure 5 shows the effective refractive index vs. confinement loss for the bundled fiber with 70μm diameter rods. Figure 6 shows the mode number vs. confinement loss (right) for the bundled fiber with 70μm diameter rods.

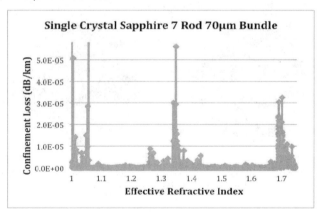

Figure 5. Effective Refractive Index vs. Confinement Loss (dB/km) for the bundled fiber with 70μm diameter rods.

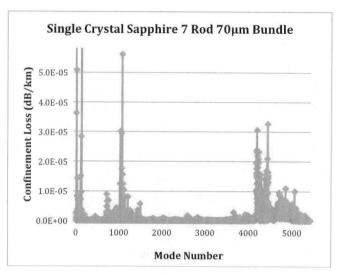

Figure 6. Mode Number vs. Confinement Loss (dB/km) for the bundled fiber with 70μm diameter rods.

Figure 7 shows the effective refractive index vs. confinement loss for the bundled fiber with 50μm diameter rods. Figure 8 shows the mode number vs. confinement loss (right) for the bundled fiber with 50μm diameter rods. The data shows that the confinement loss for the bundled fiber with 50μm diameter rods is much less than that of the bundled fiber with 70μm diameter rods. This is mainly due to the reduction of modes in the 50μm diameter rod case.

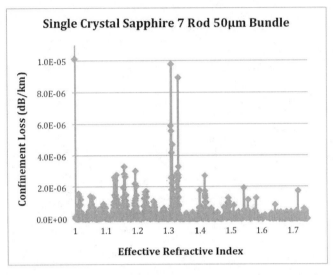

Figure 7. Effective Refractive Index vs. Confinement Loss (dB/km) for the bundled fiber with 50μm diameter rods.

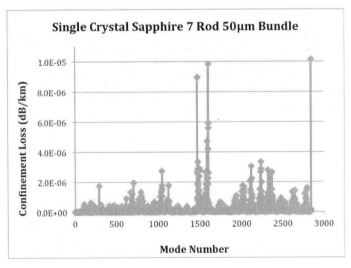

Figure 8. Mode Number vs. Confinement Loss (dB/km) for the bundled fiber with 50μm diameter rods.

CONCLUSION

Two sapphire photonic crystal fiber bundles have been presented and modeled using a multi-physics FEM modeling program. Other single crystal sapphire fiber bundles with this type of structure have been previously modeled but this is the first attempt at optimization by reducing the diameter of the core and surrounding fibers. The bundled fiber with 70μm diameter rods and the bundled fiber with 50μm diameter rods were modeled at a free space wavelength of 1.55μm with the rods all being composed of single crystal sapphire while surrounded by air. This was modeled to see if resulting changes in optical characteristics of each fiber could be determined. This process allows for easy modification of the current size and spacing of the fiber rods in order to further analyze what effect this will have on properties including confinement loss and power flow. The results also show that the confinement loss is directly related to which mode is propagating through the fiber.

ACKNOWLEDGEMENTS

The authors would like to gratefully acknowledge funding for this work from DOE NETL under contract number FC26-05NT42441.

REFERENCES

[1] K. A. Murphy, Shari Feth, Ashish M. Vengsarkar, R. O. Claus, "Sapphire fiber interferometer for microdisplacement measurements at high temperature," Proc. SPIE, vol. 1588, (Boston, MA), September 1991

[2] Nubling R. K. and J. A. Harrington. 1997. Optical properties of single-crystal sapphire fibers. *Appl. Opt.* 36:5934-5940.

[3] Desu, S. B. et. al. 1990. High temperature sapphire optical sensor fiber coatings. *SPIE Int. Soc. Opt. Eng. Proc. SPIE Int. Soc. Opt. Eng.* 1307:2-9.

[4] Neal Pfeiffenberger, Gary Pickrell, Karen Kokal and Anbo Wang, "Sapphire photonic crystal fibers", *Opt. Eng.* 49, 090501 (Sep 02, 2010); doi:10.1117/1.3483908

[5] Neal Pfeiffenberger; Gary Pickrell "Finite Element Modeling of Sapphire Photonic Crystal Fibers" Virginia Tech. MS&T , Houston, TX October 21, 2010.

[6] J. C. Knight, T. A. Birks, P. J. Russell, and J. P. de Sandro, "Properties of photonic crystal fiber and the effective index model," J. Opt. Soc. Am. 15, 748–752 (1998).

[7] D. Gloge, "Weakly guiding fibers", Appl. Opt. 10 (10), 2252 (1971)

[8] Jianming Jin, "The Finite Element Method in Electromagnetics," Second Edition, John Wiley & Sons, Inc, New York 2002.

[9] K. Saitoh and M. Koshiba, "Full-vectorial imaginary-distance beam propagation method based on finite element scheme: Application to photonic crystal fibers," IEEE J. Quantum Electron. **38**, 927-933 (2002).

[10] Kunimasa Saitoh and Masanori Koshiba, "Leakage loss and group velocity dispersion in air-core photonic bandgap fibers," Opt. Express **11**, 3100-3109 (2003)

P-TYPE SILICON OPTICAL FIBER

By Brian Scott[1]*, Ke Wang[1], Adam Floyd[1], and Gary Pickrell[1]

[*] Corresponding-Author

[1] Materials Science and Engineering Dept.
Virginia Polytechnic and State University
Blacksburg, VA 24061

ABSTRACT

Fabrication of silicon optical fibers with p-type silicon cores is experimentally demonstrated. In addition, optical and material characterizations of these fibers are presented. The fabricated fibers have outer diameters from 200 μm to 1 mm and inner diameters from 10 to 200 μm. Crystal orientation analysis using Electron Backscatter Diffraction shows that the core is polycrystalline along the fiber axis. Energy dispersive spectroscopy and secondary ion mass spectrometry analysis verifies the p-type silicon core with a boron doping level of 8.48×10^{18} /cm^3.

INTRODUCTION

Silicon is the foundation for the semi-conductor industry and is the most widely used material in the manufacture of integrated circuits and solid state devices. In addition to the use in IC manufacturing, silicon also has many uses in photonics applications, such as infrared photo detectors and mid-wavelength range waveguides. Silicon in optical fiber form will present many opportunities for increasing the capabilities of optical fibers for use as sensors and information transmission. The benefits of this type of fiber include transmission wavelengths up to 6μm which can potentially be used as Raman amplifiers or mid-IR optical sensors for their unique signatures in this range. In this paper, the fabrication and characterization of a P-type doped silicon optical fiber is demonstrated. The fiber presented is a large grain polycrystalline structure with a high retained dopant level suitable for further device fabrication. In addition to the morphological characterizatio, optical transmission in the telecommunications wavelength band is demonstrated.

Silicon core optical fibers have recently been fabricated.[1-3] However the use of these fibers is limited due to size and length produced and to the limited ability to fabricate basic electrical components without additional doping steps. In order to increase the ease of fabricating the simple components that make up more complex integrated circuits, the silicon core of the fiber should be doped to either a p-type or n-type configuration depending upon which type of circuit structure will be produced. Producing an optical fiber with this type of core will initially allow for the fabrication of simple circuits within the fiber core using standard IC fabrication methods. This in turn will lead to in-fiber opto-electronic devices which will improve signal efficiency and sensor capabilities. Silicon has low-loss transmission in the mid-wave infrared region. Incorporating this material as the core material will allow for these material characteristics to be utilized in the fiber in addition to the information transmission capabilities of typical telecommunication optical fibers.

Fabrication of semi-conductor devices is accomplished with silicon that is doped to produce extra electrons or holes in the semi-conductor. This doped silicon can then be used to build diodes, transistors and other devices through additional selective doping of regions to make these devices. Incorporation of a dopant into the silicon in the core of optical fibers may allow in fiber opto-electronic devices to be realized. This incorporation will lead to more versatile and efficient signal processing. Depending on the type of circuit or device desired either a n-type or p-type silicon core will be required. Previously an n-type silicon core fiber was fabricated and this is a continuation of that fabrication research.[4]Fabrication of the fiber was accomplished using a powder-in-tube preform, where the fibers are draw cast to produce various diameters and lengths of fiber all with a p-type silicon core.

Methods

The beginning preform that was used to fabricate these fibers is made from optical grade silica tubing manufactured by GE. A silica tube of optical purity with an OD of 6 mm and an ID of 2 mm was used with one end of the tube being closed off through the application of heat to collapse the tube. The tubing was closed off in order to hold the doped silicon powder and is done by pulling the middle of the tube to close off the center portion. This will prevent moisture contamination of the inside of the tube from the flame. A portion of the tubing was charged with the silicon powder and packed to maximize the amount of silicon in the tube. The preform was then secured into a dual chuck system with the open end of the tube attached to a vacuum system with the tube being evacuated down to a pressure of approximately 3500 mtorr. Removal of the air is thought to be a crucial step to prevent oxidation of the silicon and the formation of crystalline or amorphous silica during the fiber pulling process.

Fabrication of the fiber is done through heating the preform to temperatures higher than 1900°C, at which point the silicon has melted and the silica is sufficiently softened to allow for pulling of the fiber. After reaching the appropriate temperature the preform was pulled linearly to draw a fiber from the preform. This technique produces varying fiber diameters depending on the temperature of the silica glass and the speed at which the fiber is pulled from the preform.

Optical and morphological characterization of the fiber and the fiber core was accomplished to determine the fiber transmission, core composition, and crystallinity of the core. Transmission of the fiber was measured by aligning the lead-in and lead-out fibers with the core of the p-type fiber core and recording the spectrum. The silicon fiber was mounted on an optical stage and aligned with the lead-in and the lead-out single-mode fibers, which are mounted on separate three-dimensional stages for precision alignment. Light is delivered from the lead-in fiber into the silicon fiber. The transmitted light is collected by the lead-out fiber which transmits the signal into the detector. The optical transmission is measured in the wavelength range between 1520 nm and 1570 nm by a Micron Optics CTS system (model number SI-720).

Material characterization was done with scanning electron microscopy (SEM), energy dispersive spectroscopy (EDS), electron backscatter diffraction (EBSD) and secondary ion mass spectroscopy (SIMS). SEM samples were prepared by cleaving the fibers and mounting on samples stubs with carbon tape. EDS was performed using an accelerating voltage of 20KV and an aperture of 120μm with a count time of 90s. Samples for EBSD were mounted onto a copper wire scaffold which was then mounted into a Bakelite container for polishing. Two orientations of the fibers were prepared for analysis, the end face of the fiber and parallel to the fiber axis. Samples were polished down to a 0.3μm alumina polishing media and then polished on a vibratory polisher using a 0.05 colloidal silica suspension. Samples for the SIMS analysis were mounted in 99.99% pure indium after the end face of the fiber was polished on 1200 grit paper.

The optical transmission of the p-type silicon fiber is shown in Fig. 1, which indicates that the fibers exhibited a high transmission loss. Fig. 2 is an SEM image of the fracture surface of the end face of the fiber. The core is barely visible as a darker grey region within the lighter center fracture plateau. A elemental analysis through EDS is shown if Fig. 3. These maps show the locations of the silicon and oxygen present over the entire end face of the fiber, with the exception that no oxygen is present in the core region. Electron backscatter data is shown in Figs. 4 and 5 with the end face crystal orientation and axis crystal orientation shown respectively. The end face of the fiber shows a single crystal orientation, while the fiber view parallel to the fiber axis is decidedly poly crystalline with very large grain sizes in the core. The elliptical view of the end face is due to the large mounting angle used during sample analysis, which foreshortens the image in one direction. Figures 4 and 5 also indicated the relative crystal orientations of neighboring grains by the superimposed lattice markers in the images. Each outlined region represents a unique grain crystal orientation. Bulk concentration levels of the dopant were measured through SIMS analysis, in which it confirms the presence of boron in the core region with a concentration level of $8.48 \times 10^{18}/cm^3$.

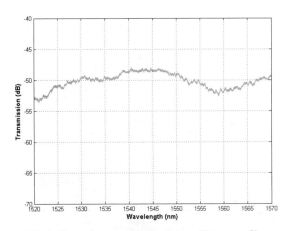

Fig. 1- Transmission spectra for P-type silicon core fiber

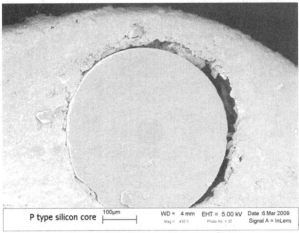

Fig. 2- SEM of P-type silicon core fiber

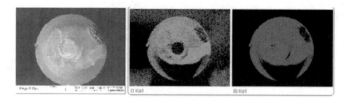

Fig. 3- Elemental mapping of P-type silicon core fiber end face

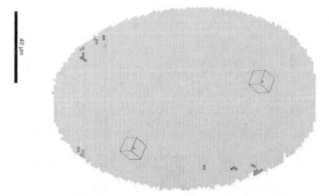

Fig. 4- EBSD Crystal orientation map of silicon core viewed from the end face

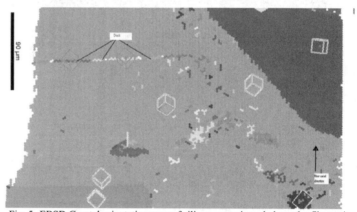

Fig. 5- EBSD Crystal orientation map of silicon core viewed along the fiber axis

Fabrication of the fiber was accomlished by a draw casting method starting with a powder in tube type preform. This type of preform allows for versatility in composition and changing of processing parameters including the core to cladding ratios that are important in developing a robust fiber. In addition to the use of a powder in tube preform, the use of a dual chuck apparatus for drawing the fibers provides additional versatility in process parameters where the preform temperature, heated regions and cladding glass viscosities can be monitored and controlled in real time as the local preform conditions dictate. This versatility allows for longer fiber lengths and varying diameters to be produced from the same preform in comparison to other fabrication methods.

This development of a p-type doped silicon core fiber may allow for the fabrication of electronic devices within the fiber core. The starting point in realizing these in fiber devices is the fabrication of a doped silicon core fiber. In this article we have demonstrated this type of fiber with a p-type doped silicon core. Elemental mapping of the fiber end-face shows that the core of the fiber is silicon with a silica cladding. Boron is not shown in the mapping due to the dopant levels (and the low atomic mass of boron) being below the detection limit of the EDS method. Confirmation of the dopant levels was done through the use of SIMS analysis. The boron dopant concentration in the silicon was determined to be $8.48 \times 10^{18} / cm^3$.

The core of this fiber is composed of a few large grains as evidence by the results from the EBSD in figures 4 and 5 which shows the end-face and longitudinal fiber sections. In the samples analyzed, the grain size of the fiber is very large with diameters running the width of the fiber. This is seen in figure 4 where the end face of the fiber appears as a single crystal. This presents the opportunity to have large regions of essentially single crystal silicon where circuits and devices can be fabricated. Further heat treatment of the fibers would promote recrystallization of the silicon core, which would eliminate the smaller grains in the core, thus reducing the effect of grain boundaries on the transmission.

The optical transmission characteristics were also studied. Fiber transmission shows a high amount of loss over the length of the fiber. The overall coupling efficiency is highly related to the optical alignment. Therefore, an accurate estimation of the transmission loss is difficult at this time. Structural characterization of the fiber core reveals that cracking occurs in the core for large diameter fibers due to the coefficient of thermal expansion mismatch. The presence of cracks will influence the transmission characteristics of the fibers. As the core to cladding ratio decreases the CTE mismatch becomes less important and the amount of cracking decreases which improve the transmission characteristics of the fiber. Therefore fibers with smaller cores will exhibit less cracking. Further structural characterization will be necessary to understand the role played by the cladding stresses that are generated during the drawing process and how this affects the solidification of the silicon.

CONCLUSION

Relatively long lengths of silicon core optical fiber has been fabricated by draw casting from a powder-in tube preform with diameters ranging from 200μm to 1mm. The silicon core is doped with boron with a dopant level of $8.48 \times 10^{18}/cm^3$, producing a p-type semi-conductor core. Crystal orientation of the fiber core is varied as the silicon forms a polycrystalline structure as it solidified from the melt after being drawn into fiber form. Transmission spectra of the fiber in the 1520-1570 nm range show s a relatively high loss. This loss is potentially due to the free space connections used in the transmission analysis, roughness of the fiber endface, and the variations present in the core.

REFERENCES
[1]   P. J. A. Sazio, *et al.*, "Microstructured optical fibers as high-pressure microfluidic reactors," *Science* vol. **311**, pp. 1583-1586, 2006.
[2]   J. Ballato, *et al.*, "Silicon optical fiber," *Optics Express* vol. **16**, pp. 18675-18683, 2008.
[3]   B. Scott, *et al.*, "Fabrication of silicon optical fiber," *Optical Engineering Letters,* vol. 48, 2009.
[4]   B. Scott, *et al.*, "Fabrication of N-type Silicon Optical Fibers," *Photonics Technology Letters,* vol. 21, pp. 1798-1800, 2009.

SYNTHESIS AND CHARACTERIZATION OF COBALT ALUMINATE AND Fe_2O_3
NANOCOMPOSITE ELECTRODE FOR SOLAR DRIVEN WATER SPLITTING TO PRODUCE
HYDROGEN

Sudhakar Shet,[1, 2,] Kwang-Soon Ahn ,[3] Yanfa Yan,[1] Heli Wang,[1] Nuggehalli Ravindra,[2] John Turner,[1]
and Mowafak Al-Jassim[1]

[1] National Renewable Energy Laboratory, Golden, CO 80401 USA
[2] New Jersey Institute of Technology, Newark, NJ 07102 USA
[3] School of Display and Chemical Engineering, Yeungnam University, Gyeongsan, 712-
749, S. Korea

ABSTRACT
    Cobalt aluminate-Fe_2O_3 p-n nanocomposite electrodes were deposited on the silver coated stainless
steels and annealed at 800 °C. Their photoresponses were investigated and compared with that of p-
type Cobalt aluminate. We found that nanocomposite electrodes exhibited much improved
photoresponses as compared to p-type Cobalt aluminate. We attribute the improvement to the band
offset at the formed three-dimensional p-n junction, which promote photo-generated carrier separation
and reduce carrier recombination.

INTRODUCTION
    Transition metal oxide-based photoelectrochemical (PEC) splitting of water has attracted wide
interest since the discovery of photoinduced decomposition of water on TiO_2 electrodes.[1-9] To date,
most investigations have focused on n-type materials such as TiO_2, ZnO, WO_3, Fe_2O_3, *etc* due to the
potential stability in aquesou solutions.[1-9] For the application of water splitting, the use of both n-type
and p-type semiconductors is often desirable,[10, 11] because hydrogen is evolved from the p-side and
oxygen from the n-side electrodes. Unfortunately, most p-type materials (Cu_2O, CdSe, MoS_2, *etc*.)
developed so far are very weak against photocorrosion.[11] Recently, Parkinson's group [12, 13] and our
group [14] have reported experimentally and theoretically that the Co-based spinel oxides such as
$CoAl_2O_4$ are possible candidates as p-type electrodes for PEC water splitting, because these p-type
oxides are found to be very stable in aqueous solution. However, the photo-responses of these oxides
were found to be weak.[12,13] It is therefore necessary to develop approaches to enhance their photoresponses. Like other nanostructures, nanoparticles are often used an approach to enhance PEC
response due to the greatly increased surface areas. However, unlike thin film electrodes, nanoparticles
can hardly develop depletion layer due to the small size when they contact with electrolytes. The
electric field generated in the depletion layer is generally beneficial for photovoltaic devices as it help
to separate photogenerated electron-hole pairs and reduces carrier recombination.[15,16] Thus, attempts to
suppress recombination rate in the nanostructures have been performed by energy band engineering or
developing p-n nanojunction.[17]
    In this paper, we develop p-n nanocomposite electrodes consisting of p-$CoAl_2O_4$ and n-Fe_2O_3
nanoparticles for enhancing PEC photo-response. The performance of these p-n nanocomposite
electrodes is compared with that of sole p-$CoAl_2O_4$ nanoparticles electrodes. All synthesized
electrodes exhibited p-type PEC responses and were very resistant against photocorrosion. However,
the nanocomposite electrodes exhibited much improved PEC photo-response as compared to the
reference p-type $CoAl_2O_4$ electrodes. We speculate that the enhancement is due to the formation of
three dimensional junction between p-type $CoAl_2O_4$ and n-type Fe_2O_3 nanoparticles, which promotes
electron-hole separation and thus reduces charge recombination. Our results suggest that three

dimensional p-n junction may be considered for enhancing PEC photo-response for nanostructure electrodes.

EXPERIMENTAL

The preparation of $CoAl_2O_4$-Fe_2O_3 p-n nanocomposite film electrodes starts from dispersing $CoAl_2O_4$ and Fe_2O_3 nanoparticles (size < 50 nm, Sigma-Aldrich Co.) in ethanol by paint shaking for 2 hr. Mixed nanoparticles with Fe_2O_3 concentration from 5 wt% to 20 wt% were prepared. These colloids were thoroughly dispersed using a conditioning mixer by adding ethyl cellulose as the binder and α-terpineol as a solvent for the pastes, followed by concentration using an evaporator. The pastes were doctor-bladed on Ag coated stainless steel substrates (Ag/SS) and followed by calcinations at 800 °C for 4 hrs in air to remove the binder. All samples have a similar film thickness of about 6 μm as measured by stylus profilometry.

The structural and crystallinity characterizations were performed by X-ray diffraction (XRD) measurements using an X-ray diffractometer. The surface morphology was examined by field-emission scanning electron microscopy (FE-SEM). PEC measurements were performed in a three-electrode cell with a flat quartz window to facilitate illumination of the photoelectrode surface.[18-26] The nanocomposite films and $CoAl_2O_4$ nanoparticel films (active area: 0.25 cm^2) were used as the working electrodes. A Pt sheet (area: 10 cm^2) and a Ag/AgCl electrode (with saturated KCl) were used as counter and reference electrodes, respectively.[27-33] A 0.5-M NaOH basic aqueous solution (pH ~ 13) was used as the electrolyte. The PEC response was measured using a fiber-optic illuminator (150-W tungsten-halogen lamp) with a UV/IR cut-off filter (cut-off wavelengths: 350 and 750 nm). Light intensity with a UV/IR filter was 75 mW/cm^2 and was measured by a photodiode power meter. The PEC response was measured with a time under the light on/off illumination at constant potential. Photocurrent behavior with single wavelength was also measured at -1 V at room temperature using a monochromator, in which the steady-state values of photocurrents at each wavelength were collected after 2 min stabilization and were corrected for dark currents.

RESULTS AND DISCUSSION

Figure 1 shows X-ray diffraction patterns for SS and Ag/SS before and after annealing at 800 °C in air. The SS shows the formation of iron oxide (* peaks) on the surface after the annealing, indicating that SS is not appropriate as the substrate for $CoAl_2O_4$ electrodes, because iron oxide has very poor electrical conductivity, which makes current collection from the $CoAl_2O_4$ to the SS difficult. On the contrary, Ag/SS exhibited no evidence of formation of oxides after the annealing at 800 °C in air. Therefore, $CoAl_2O_4$-Fe_2O_3 nanocomposite films could be coated on Ag/SS and annealed at temperature of 800 °C without substrate deterioration (See the XRD curve of the annealed $CoAl_2O_4$/Ag/SS sample in Fig. 1).

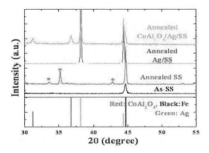

Fig. 1. XRD curve of unannealed and annealed SS substrates, annealed Ag/SS substrate and $CoAl_2O_4$/Ag/SS.

The SEM image shown in Figure 2 indicates that the annealed nanocomposite electrode is nanoporous and its particle size is also well corresponding to the crystallite size (33 nm) calculated from the XRD peak around 36.8 °. The particle size is the same as the unannealed particles, indicating that on obvious solid reaction occurred during the annealing.

Fig. 2. SEM image of annealed $CoAl_2O_4$/Ag/SS

The measured photocurrent as a function of potential is shown in Fig. 3. It shows that the on-set potential of the photocurrent is at -0.2 V and the photocurrent saturates from -0.6 V. The inset in Fig. 3 shows the photocorrosion stability during the cell operation at -1 V. It is seen that $CoAl_2O_4$ is very stable in the basic aqueous solution, a property that not typically seen for p-type materials. The PEC response measured for pure $CoAl_2O_4$ nanoparticle electrode under the light on/off conditions, photocurrent was generated in *cathodic* direction (not shown here), indicating that $CoAl_2O_4$ is p-type, which is due to the native cation vacancies as suggested by our theoretical study.[14]

$CoAl_2O_4$ (band gap about 2.3 eV) is intrinsically p-type. On contrary, Fe_2O_3 (band gap about 2.0 eV) is inherently n-type, due to oxygen vacancies.[34] When p-type $CoAl_2O_4$ and n-type Fe_2O_3 particles are interconnected, a three-dimensional p-n junction would form and their valance bands (VBs) and conduction bands (CBs) will have offset in a way that the VB and CB of p-type $CoAl_2O_4$ is higher in energy than that of n-type Fe_2O_3 due to Fermi level alignment.

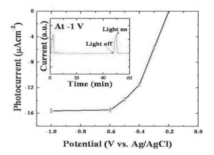

Fig. 3. Measured I-V curve for pure CoAl$_2$O$_4$ nanoparticle electrode.

Unlike bulk p-n junction, the three-dimensional junction may not have clear depletion region due to the nanoparticle size. Jonetheless, such ba nd offset at the junction promotes photo-generated electron-hole pair separation and reduce carrier recombination, and therefore could lead to enhanced PEC response. The p-n nanocomposite electrodes were prepared by mixing the different amount of Fe$_2$O$_3$ nanoparticles (5 to 20 wt%) in the CoAl$_2$O$_4$ nanoparticles.

Figure 4 shows the comparison of PEC responses of a nanocomposite film with 5 wt% Fe$_2$O$_3$ and a reference CoAl$_2$O$_4$ nanoparticle film. The photo-current is generated in cathodic direction, meaning that the overall electrode behaves like p-type. The saturated photo-currents are lined up for comparison. It shows clearly that the photo-current generated by the nanocomposite film is much larger than that generated by sole p-type CoAl$_2$O$_4$ nanoparticle film.

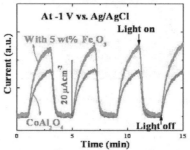

Fig. 4. Comparison of PEC responses measured for pure CoAl$_2$O$_4$ nanoparticle and p-n nanocomposite electrodes with a time under the light on/off conditions at constant -1 V.

Figure 5 shows the photocurrents measured at constant -1 V for nanocomposite films with different amount of Fe$_2$O$_3$. It shows that all nanocomposite CoAl$_2$O$_4$-Fe$_2$O$_3$ p-n nanocomposite films exhibit much improved PEC responses than the sole CoAl$_2$O$_4$ nanoparticle film. It is also seen that the enhancement does not increase linearly as the increase of the amount of Fe$_2$O$_3$ nanoparticles, because too much Fe$_2$O$_3$ would indicate less amount of p-type CoAl$_2$O$_4$ and less contact area with electrolyte.

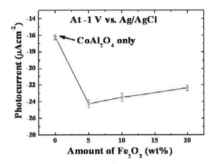

Fig. 5. Photocurrents measured at constant -1 V for nanocomposite films with different amount of Fe_2O_3.

Photocurrent measured at -1 V as a function as wavelength for the $CoAl_2O_4$ with 10 wt% Fe_2O_3, showed that the photoresponse of the nanocomposite film is occurred only at the wavelengths less than 532 nm (2.33 eV), which corresponds to the band gap of $CoAl_2O_4$ rather than that of Fe_2O_3. This result further indicate that the enhanced photoresponses of nanocomposite films are not due to the contribution from Fe_2O_3, but due to the reduced carrier recombination or carrier separation promoted by the three-dimensional p-n junction. It should be noted that both $CoAl_2O_4$ and nanocomposite electrodes exhibited slow carrier transport kinetics due to the large effective mass for both electrons and holes in $CoAl_2O_4$.

CONCLUSIONS

Cobalt aluminate-Fe_2O_3 p-n nanocomposite electrodes were deposited on the silver coated stainless steels and post annealed at 800 °C. Their photoresponses were investigated and compared with that of p-type Cobalt aluminate. We found that nanocomposite electrodes exhibited much improved photoresponses as compared to p-type Cobalt aluminate. We attribute the improvement to the band offset at the formed three-dimensional p-n junction, which promote photo-generated carrier separation and reduce carrier recombination. Our results suggest that three dimensional p-n junction may be considered for enhancing PEC photo-response for nanostructure electrodes.

REFERENCES
[1] A. Fujishima and K. Honda, *Nature* 238, p. 37, 1972.
[2] R. Asahi, T. Morikawa, T. Ohwaki, K. Aoki, and Y. Taga, *Science* 293, p. 269, 2001.
[3] O. Khaselev and J. A. Turner, *Science* 280, p. 425, 1998.
[4] V.M. Aroutiounian, V.M. Arakelyan, and G.E. Shahnazaryan, *Solar Energy* 78, p. 581, 2005.
[5] J. Yuan, M. Chen, J. Shi, and W. Shangguan, *Inter. J. Hydrogen Energy*, 31, p. 1326, 2006.
[6] G. K. Mor, K. Shankar, M. Paulose, O. K. Varghese, and C. A. Grimes, *Nano Lett.* 5, p. 191, 2005.
[7] B. O'Regan and M. Grätzel, *Nature* 353, p. 737, 1991.
[8] K. Kakiuchi, E. Hosono, and S. Fujihara, *J. Photochem. & Photobiol. A: Chem.* 179, p. 81, 2006.
[9] T. F. Jaramillo, S. H. Baeck, A. Kleiman-Shwarsctein, and E. W. McFarland,

Macromol. Rapid Comm. 25, p. 297, 2004.

[10] A. J. Įozik, Appl. Phys. Lett. 29, p. 150, 1976.

[11] G. K. Mor, O. K. Varghese, R. H. T. Wilke, S. Sharma, K. Shankar, T. J. Latempa, K. –S. Choi, and C. A. Grimes, *Įano Lett* . 8(7), p. 1906, 2008.

[12] M. Woodhouse, G. S. Herman, and B. A. Parkinson, *Chem. Mater.* 17, p. 4318, 2005.

[13] M. Woodhouse and B. A. Parkinson, *Chem. Mater.* 20, p. 2495, 2008.

[14] A. Walsh, S. –H. Wei, Y. Yan, M. M. Al-Jassim, and J. A. Turner, *Phys. Rev. B* 76, p. 165119, 2007.

[15] G. Schlichthrl, S. Y. Huang, J. Sprague, and A. J. Frank, *J. Phys. Chem. B* 101(41), p. 8141, 1997.

[16] R. Beranek, H. Tsuchiya, T. Sugishima, J.M. Macak, L. Taveira, S. Fujimoto, H. Kisch, and P. Schmuki, *Appl. Phys. Lett.* 87, p. 243114, 2005.

[17] H. G. Kim, P. H. Borse, W. Choi, and J. S. Lee, *Angew. Chem. Int. Ed.* 44, p. 4585 ,(2005).

[18] S. Shet, K. –S. Ahn, T. Deutsch, H. Wang, Į. Ravindra, Y. Yan, J. Turner, M. Al-Jassim, *J. Mater. Research* 25, 69 Doi: 10.1557/JMR.2010.0017, 2010.

[19] K.–S. Ahn, Y. Yan, S. Shet, T. Deutsch, J. Turner, and M. Al-Jassim, *Appl. Phys. Lett.* 91, p. 231909, 2007.

[20] S. Shet, K. –S. Ahn, H. Wang, Į. Ravindr a, Y. Yan, J. Turner, M. Al-Jassim, *J. Mater. Science* DOI 10.1007/s10853-010-4561-x, 2010.

[21] S. Shet, K. –S. Ahn, Y. Yan, T. Deutsch, K. M. Chrusrowski, J. Turner, M. Al-Jassim, and Į. Ravindra, *J. Appl. Phys.* 103, p. 073504, 2008.

[22] K. –S. Ahn, S. Shet, T. Deutsch, C. S. Jiang, Y. Yan, M. Al-Jassim, and J. Turner, *J. Power Source*, 176, p. 387, 2008.

[23] S. Shet, K.-S. Ahn, T. Deutsch, H. Wang, Į. Ravindra, Y. Yan, J. Turner, M. Al-Jassim, *J. Power Sources* 195, p. 5801, 2010.

[24] K.-S. Ahn, Y. Yan, S. Shet, K. Jones, T. Deutsch, J. Turner, M. Al-Jassim, *Appl. Phys. Lett.* 93, p. 163117, 2008.

[25] S. Shet, K. –S. Ahn, Į. Ravindra, Y. Yan, J. Turner, M. Al-Jassim, *J. Materials* 62, p. 25, 2010.

[26] S. Shet , K. Ahn, Į. Ravindra, Y. Ya n, T. Deutsch, J. Turner, M. Al-Jassim *Proceedings of the Materials Science & Technology*, p. 219, 2009.

[27] S. Shet, K. Ahn, Į. Ravindra, Y. Yan, T. Deutsch, J. Turner, M. Al-Jassim *Proceedings of the Materials Science & Technology*, p. 277, 2009.

[28] Y. Yan, K. Ahn, S. Shet, T. Deutsch, M. Huda, S. Wei, J. Turner, M. Al-Jassim, *Proceedings of the SPIE*, 6650, p. 66500H, 2007.

[29] S. Shet, K. Ahn, Į. Ravindra, Y. Yan, T. De utsch, J. Turner, M. Al-Jassim, Materials Science & Technology 2009, *Ceramic Transactions volume*, (2010) in press.

[30] K.-S. Ahn, S. Shet, Y. Yan, J. Turner, M. Al-Jassim, Į. M. Ravindra, *Proceedings of the Materials Science & Technology*, p. 901, 2008.

[31] S. Shet, K. –S. Ahn, Į. Ravindra, Y. Yan, J. Turner, M. Al-Jassim, *J. Materials* 62, p. 25, 2010.

[32] K. Ahn, S. Shet, T. Deutsch, Y. Yan, J. Turner, M. Al-Jassim, Į. M. Ravindra, *Proceedings of the Materials Science & Technology*, p. 952, 2008.

[33] S. Shet, K. Ahn, T. Deutsch, Y. Yan, J. Turner, M. Al-Jassim, Į. Ravindra, *Proceedings of the Materials Science & Technology*, p. 920, 2008.

[34] Y. –S. Hu, A. Kleiman-Schwarsctein, A. J. Forman, D. Hazen, J. –Į. Park, and E. W. McFarland, *Chem. Mater.* 20(12), p. 3803, 2008.

INFLUENCE OF SUBSTRATE TEMPERATURE AND RF POWER ON THE FORMATION OF ZnO NANORODS FOR SOLAR DRIVEN HYDROGEN PRODUCTION

Sudhakar Shet,[1, 2,] Heli Wang,[1] Yanfa Yan,[1] Nuggehalli Ravindra,[2] John Turner,[1] and Mowafak Al-Jassim[1]

[1] National Renewable Energy Laboratory, Golden, CO 80401 USA
[2] New Jersey Institute of Technology, Newark, NJ 07102 USA

ABSTRACT

We report on the effects of substrate temperature and RF power on the formation of aligned nanorods in ZnO thin films. ZnO thin films were sputter-deposited in mixed Ar and N_2 gas ambient at various substrate temperatures and RF power. At low substrate temperatures (below 300 °C), ZnO nanorods do not form regardless of RF power. High RF power helps to promote the formation of aligned ZnO nanorods. However, lower RF power usually lead to ZnO films with better crystallinity at the same substrate temperatures in mixed Ar and N_2 gas ambient and therefore better photoelectrochemical (PEC) response.

INTRODUCTION

Photoelectrochemical (PEC) systems based on transition metal oxides, such as TiO_2, ZnO, and WO_3, have received extensive attention since the discovery of photoinduced decomposition of water on TiO_2 electrodes.[1-3] ZnO has similar bandgap (~3.3 eV) and band-edge positions as compared to TiO_2. ZnO has a direct bandgap and higher electron mobility than TiO_2.[2] Thus ZnO could also be a potential candidate for PEC splitting of water for H_2 production.[3]

To improve PEC performance, a photoelectrode should have a high contact area with the electrolyte to provide more interfacial reaction sites. Therefore, the morphological features of the thin films, such as grain size, grain shape, and surface area would have profound influence on the performance of thin film electrodes for PEC applications. It has been expected that electrodes with nanostructures would exhibit improved PEC performance as compared to those without nanostructures. In our earlier studies[4], we have found that aligned single crystal ZnO nanorods along the c-axis can be synthesized by RF-sputter-deposition in mixed Ar and N_2 ambient, and ZnO films with aligned nanorods exhibited improved performance as compared to ZnO films without nanorods. So far, RF sputtering is much less considered than other methods for the growth of ZnO nanorods. Detailed examination on this method for ZnO nanorod growth is needed. In this paper, we report on effects of substrate temperatures and RF power on the formation of aligned nanorods in ZnO thin films.

EXPERIMENTAL

The $ZnO(Ar:N_2)$ films were deposited by reactive RF sputtering a ZnO target using an Argon/Nitrogen gas mixture. Transparent conducting F-doped SnO_2 (FTO) coated glass (20–23 $\Omega/\square$) were used as the substrate to allow PEC measurements. The detailed growth conditions can be found in our earlier publications. ZnO films were deposited at RF power of 100W to 300W and substrate temperature 100 to 500°C. All the deposited samples were controlled to have similar film thickness of 0.5 for 100W and 1 μm for 200 and 300W as measured by stylus profilometry.

The structural and crystallinity characterizations were performed by X-ray diffraction (XRD) measurements, using an X-ray diffractometer (XGEN-4000, SCINTAG Inc.), operated with a Cu Kα radiation source at 45 kV and 37 mA. The N concentration in the $ZnO(Ar:N_2)$ films was evaluated by X-ray photoelectron spectroscopy (XPS). The surface morphology was examined by atomic force microscopy (AFM) conducted in the tapping mode with a silicon tip, and field emission scanning electron microscopy (FE-SEM). The UV-Vis absorption spectra of the samples were measured by an n&k analyzer 1280 (n&k Technology, Inc.) to investigate the optical properties.

PEC measurements were performed in a three-electrode cell with a flat quartz-glass window to facilitate illumination to the photoelectrode surface.[5-14] The sputter-deposited films were used as the working electrodes. Pt plate and an Ag/AgCl electrode were used as counter and reference electrodes, respectively. A 0.5-M Na_2SO_4 mild aqueous solution was used as the electrolyte for the stability of the ZnO.[15-20] Photoelectrochemical response was measured using a fiber optic illuminator (150 W tungsten-halogen lamps) with a UV/IR filter. Light intensity was measured by a photodiode power meter, in which total light intensity with the UV/IR filter was fixed to 125 mW/cm^2.

RESULTS AND DISCUSSION

We first present the results of ZnO thin films grown at 100 W RF power and different temperatures (from 100 to 500 °C). The X-ray diffraction curves obtained from these films are shown in Figures 1(a). No clear preferred orientation was observed for samples deposited at temperatures below 400 °C, indicating that no obvious formation of aligned nanorods at these temperatures when 100 W RF power was applied. However, for samples deposited above 400 °C, some degree of preferred orientation is observed. The crystallinity of these films increases gradually with the increase of substrate temperatures. When the RF power was increased to 200 W, clear preferred orientation was observed at substrate temperatures of 400 and 500 °C as shown in Fig. 1(b). Figure 1(c) shows X-ray diffraction curves for the ZnO(Ar:N_2) films deposited at 300W RF power. The preferred orientation is also very clear for samples deposited at 400 and 500 °C.

The FWHM values of (002) peaks are larger for ZnO(Ar:N_2) films deposited at temperatures below 300°C than that grown above 300°C with the same RF power. It is also seen that at the substrate temperatures below 300 °C, the higher RF power lead to larger FWHM values. This is because that N can be incorporated at substrate temperatures below 300 °C and the incorporation of N leads to reduced crystallinity. Higher RF power leads to more N-incorporation and therefore larger FWHM values.

At substrate temperature above 300 °C, no clear N was incorporated. In these cases, the crystallinity is independent to the RF power. The N concentrations (at.%) for the ZnO(Ar:N_2) films measured by XPS showed that with the increase of substrate temperature, the N concentration decreased rapidly and disappeared at temperatures above 300°C. The preferred orientation observed from X-ray diffraction gives an indication of the formation of nanorods.

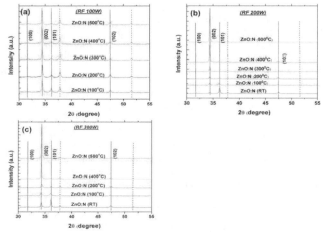

Fig. 1. X-ray diffraction curves for (a) ZnO(Ar:N$_2$)(100W) films (b) ZnO(Ar:N$_2$)(200W) films and (c) ZnO(Ar:N$_2$)(300W) films deposited at different substrate temperatures.

AFM and SEM surface imaging were carried out to verify the formation of nanorods. As an example, figure 2 shows AFM surface morphology (5×5 μm^2) of ZnO(Ar:N$_2$)(200W) films deposited at the various substrate temperatures. It shows clearly that the ZnO(Ar:N$_2$) deposited at 100°C has a random orientation. As substrate temperature increases, aligned nanorods along the c-axis were promoted to form. At 500°C, the ZnO(Ar:N$_2$) film reveals growth of hexagonallike nanorod structure.

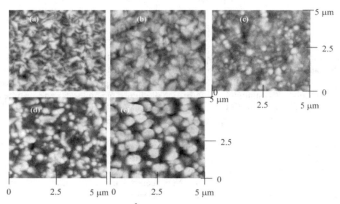

Fig. 2. AFM surface morphology (5×5 μm^2) of (a-e) the ZnO(Ar:N$_2$)(200W) films deposited at the substrate temperatures of 100, 200, 300, 400, and 500°C, respectively.

AFM images reveal that the significantly increased (002) peak in the X-ray diffraction curve obtained in ZnO(Ar:N$_2$) at different RF power is largely due to the formation of aligned nanorods along the c-axis. FE-SEM top-view image (not shown here) also confirmed vertically aligned, single crystal hexagonallike nanorods with flat (0002) surfaces in these ZnO(Ar:N$_2$) films at 500 °C with RF power of 200 W and 300 W, respectively. For ZnO(Ar:N$_2$) film deposited at 400 °C with RF power of 100 W, showed the formation of mixed pyramid-like ZnO and nanorod ZnO, indicating that the substrate temperature is critical for the formation of ZnO nanorods. The nanorod structures provide high surface areas and superior carrier transport (or conductivity) along the c-axis, which may lead to increased interfacial reaction sites and the reduced recombination rate.[11-12]

The PEC response for the ZnO(Ar:N$_2$) thin films deposited at various substrate temperature and RF power was also investigated. We found that ZnO samples with aligned nanorods indeed exhibited higher photo-currents. It is seen that the ZnO(Ar/N$_2$) film deposited at 500°C (with nanorods) exhibited much higher photocurrents than the film deposited at the lower substrate temperature (without nanorods). To see the effects of substrate temperature and RF power on PEC response, we measured photocurrent at 1.2 V potential for ZnO(Ar:N$_2$) films under continuous illumination with UV/IR filter. Figure 3 shows the measured photocurrents as a function of the substrate temperature for the ZnO(Ar:N$_2$) films at various substrate temperatures and RF power.

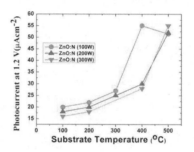

Fig. 3. Photocurrents measured at 1.2 V as a function of the substrate temperature for the ZnO(Ar:N$_2$) films deposited at different RF power.

It is seen that the substrate temperature plays an important role in photo-current. For a given RF power, higher substrate temperatures lead to improved photocurrents. ZnO(Ar:N$_2$) films deposited at 400°C for 100W and 500°C for 200 and 300W exhibits the best photoelectrochemical response. This is because of the improved crystallinity and formation of aligned nanorods at high substrate temperatures. Thus, the rapid enhancement in PEC response of the ZnO(Ar:N$_2$) films is consistent with the XRD, AFM, and FE-SEM results indicating either increased crystallinity and formation of nanorods along the c-axis.

We now discuss the results of samples prepared at substrate temperature of 100°C with RF power from 100 to 500W. Figure 4 shows X-ray diffraction curves obtained from these ZnO(Ar:N$_2$) films. The crystallinity of ZnO films decreases gradually with the increase of RF power. The measured FWHM values correlate very well with the N concentration and RF power, i.e., when RF power increases, FWHM value and N concentration also increase. This is a clear evidence of that higher RF power would enhanced the incorporation of N at low substrate temperature.

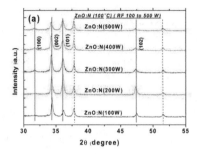

Fig. 4. X-ray diffraction curves for ZnO(Ar:N$_2$)(100°C) at RF power from 100W to 500W.

Figure 5 shows the measured photocurrents as a function of the RF power for the ZnO(Ar:N$_2$) films deposited at a substrate temperature of 100°C. It is seen that the photocurrent decreases as the RF power increases from 100 to 500W. This trend can be attributed to the decrease of film crystallinity. Therefore, high RF power is not favorable for crystallinity of ZnO films when N is available in the growth chamber.

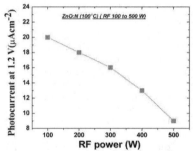

Fig. 5. Photocurrents measured at 1.2 V as a function of RF power for ZnO(Ar:N$_2$) films deposited at substrate temperature of 100°C.

CONCLUSIONS

We have synthesized and characterized ZnO thin films deposited at various substrate temperature and different RF power in mixed Ar and N$_2$ gas ambient. We found that high substrate temperature and high RF power help to promote the formation of aligned ZnO nanorods thin film grown in mixed Ar and N$_2$ gas ambient, resulting in the significantly enhanced PEC response. However, substrate temperature plays a more important role than RF power in the formation of ZnO nanorods. Our results suggest that the formation of aligned nanorod can be optimized by tuning the growth conditions, such as substrate temperatures and RF power.

REFERENCES

[1] A. Fujishima and K. Honda, *Nature* (London) 238, p. 37, 1972.

[2] K. Kakiuchi, E. Hosono, and S. Fujihara, *J. Photochem. & Photobiol. A: Chem.* 179, p. 81, 2006.

[3] T. F. Jaramillo, S. H. Baeck, A. Kleiman-Shwarsctein, and E. W. McFarland, *Macromol. Rapid Comm.* 25, p. 297, 2004.

[4] K. –S. Ahn, S. Shet, T. Deutsch, C. S. Jiang, Y. Yan, M. Al-Jassim, and J. Turner, *J. Power Source*, 176, p. 387, 2008.

[5] S. Shet, K.-S. Ahn, T. Deutsch, H. Wang, N. Ravindra, Y. Yan, J. Turner, M. Al-Jassim, *J. Power Sources* 195, p. 5801, 2010.

[6] K.–S. Ahn, Y. Yan, S. Shet, T. Deutsch, J. Turner, and M. Al-Jassim, *Appl. Phys. Lett.* 91, p. 231909, 2007.

[7] S. Shet, K. –S. Ahn, H. Wang, N. Ravindra, Y. Yan, J. Turner, M. Al-Jassim, *J. Mater. Science* DOI 10.1007/s10853-010-4561-x, 2010.

[8] K.-S. Ahn, Y. Yan, S. Shet, K. Jones, T. Deutsch, J. Turner, M. Al-Jassim, *Appl. Phys. Lett.* 93, p. 163117, 2008.

[9] H. Wang, T. Deutsch, S. Shet, K. Ahn, Y. Yan, M. Al-Jassim and J. Turner, *Solar Hydrogen and Nanotechnology IV, SPIE*, Nanoscience + Engineering, p. 7408, 2009.

[10] S. Shet , K. Ahn, N. Ravindra, Y. Yan, T. Deutsch, J. Turner, M. Al-Jassim *Proceedings of the Materials Science & Technology*, p. 219, 2009.

[11] S. Shet, K. Ahn, N. Ravindra, Y. Yan, T. Deutsch, J. Turner, M. Al-Jassim *Proceedings of the Materials Science & Technology*, p. 277, 2009.

[12] K.-S. Ahn, Y. Yan, M.-S. Kang, J.-Y. Kim, S. Shet, H. Wang, J. Turner, and M. Al-Jassim, *Appl. Phys. Lett.* 95 p. 022116, 2009.

[13] Y. Yan, K. Ahn, S. Shet, T. Deutsch, M. Huda, S. Wei, J. Turner, M. Al-Jassim, *Proceedings of the SPIE*, 6650, p. 66500H, 2007.

[14] S. Shet, K. Ahn, N. Ravindra, Y. Yan, T. Deutsch, J. Turner, M. Al-Jassim, Materials Science & Technology 2009, *Ceramic Transactions volume*, (2010) in press

[15] K.-S. Ahn, S. Shet, Y. Yan, J. Turner, M. Al-Jassim, N. M. Ravindra, *Proceedings of the Materials Science & Technology*, p. 901, 2008.

[16] S. Shet, K. –S. Ahn, N. Ravindra, Y. Yan, J. Turner, M. Al-Jassim, *J. Materials* 62, p. 25, 2010.

[17] K. Ahn, S. Shet, T. Deutsch, Y. Yan, J. Turner, M. Al-Jassim, N. M. Ravindra, *Proceedings of the Materials Science & Technology*, p. 952, 2008.

[18] S. Shet, K. Ahn, T. Deutsch, Y. Yan, J. Turner, M. Al-Jassim, N. Ravindra, *Proceedings of the Materials Science & Technology,* p. 920, 2008.

[19] S. Shet, K. –S. Ahn, Y. Yan, T. Deutsch, K. M. Chrusrowski, J. Turner, M. Al-Jassim, and N. Ravindra, *J. Appl. Phys.* 103, p. 073504, 2008.

[20] S. Shet, K. –S. Ahn, T. Deutsch, H. Wang, N. Ravindra, Y. Yan, J. Turner, M. Al-Jassim, *J. Mater. Research* 25, 69 Doi: 10.1557/JMR.2010.0017, 2010.

[21] C.M. López and K.S. Choi, *Chem. Commun.* p. 3328, 2005.

[22] M. Law, L. E. Greene, J. C. Johnson, R. Saykally, and P. Yang, *Nature Mater.* 4, p. 455, 2005.

POROUS MATERIAL FABRICATION USING ICE PARTICLES AS A PORE FORMING AGENT

Samantha Smith and Gary Pickrell
Virginia Polytechnic Institute and State University
Blacksburg, Virginia, United States

ABSTRACT
Nanoporous materials show promise in several energy related applications including catalytic membranes, adsorbent structures for hydrogen storage, and fuel cell electrodes. This study introduces the use of ice particles as a pore forming agent to fabricate porous materials. This novel method possesses several advantages over current industrial techniques including environmental friendliness, low cost, and flexibility in size and shape of resulting pores. Porous ceramic structures were created by adding preformed ice particles to an alumina slurry which was quickly frozen, air dried, and then sintered. Porosity was characterized using optical microscopy, Scanning Electron Microscopy (SEM) and Archimedes measurements. Amount of porosity was controlled through specifying the amount of ice added to the ceramic slurry. Samples were prepared with porosity levels ranging from 30-75%. Currently, observed pores are in the micron range, however this method can be extended to create a nanoporous structure.

INTRODUCTION
The field of porous ceramics has been growing rapidly as these materials have been successfully utilized in increasingly diverse and advanced technological applications. These applications include molten metal filters [1], hot gas filters [2], catalytic supports [3], thermal insulation [4], lightweight structural components [5], and bone regenerations scaffolds [6], just to name a few. Each application requires a unique set of pore structures and some of the more high tech applications can require pore structures that are quite complex. The amount of open and closed porosity, pore size and shape, pore uniformity, and degree of pore interconnectivity all play crucial roles in determining the degree of success of a porous component in its intended application.

Specific pore requirements are met through the selection and control of an appropriate ceramic processing method. Several different methods are used to manufacture porous materials, each with its own set of advantages and disadvantages. The majority of processing methods that currently exist are limited in the types of pores they can produce. Freeze-casting, for example, an environmentally friendly technique; however, this method is only able to make unidirectionally aligned cylindrical pores. An ideal processing route to porous ceramics would be able to create any type of porous microstructure within any system. It would also be easy to control, cost effective, easily performed in a manufacturing environment, and environmentally friendly. Such a technique does not currently exist, although extensive research is ongoing to identify one.

The purpose of the present study is to introduce and prove the feasibility of a novel processing route to porous materials using ice particles as a pore-forming agent (PFA) or a pore template. Ice particles can act as a pore holder when introduced to a ceramic slip. The structure can be set through freezing and a simple drying process removes the ice, leaving a void in its place. This method has the potential to meet all of the above described ideal requirements: flexibility in obtainable pore microstructures, cost effectiveness, environmental friendliness, and ability to be performed in a manufacturing environment. Flexibility in pore size and shape is easily obtained with this method. Ice particles can be created ahead of time to the exact size and shape of the desired pores within the final structure. Spherical ice particles in a wide range of sizes can be created to form uniform spherical pores in a ceramic component. If open directionally aligned tube shaped pores are desired, columnar ice crystals can be grown and held in place while a ceramic slip is poured around them. Once the structure is set and the part is dried, the directionally aligned tubular voids will be present where the ice crystals once existed. Higher degrees of porosity can be obtained by increasing the amount of ice introduced to the ceramic slip.

Ice particles have considerable advantages over typical templating materials. Most often, organic PFA's are chosen to hold the place of pores within a slip. Many different kinds of organic PFAs have been used to make porous materials including spherical polystyrene beads [7], organic polymer sponges [8], biological agents [9], poppy seeds [10], wheat flour [11], graphite [12] and polyurethane foam [13]. These organics must be burned out for removal from the system, and organic burnout poses several environmental concerns, such as the release of volatile organic components (VOCs) into the atmosphere. Ice, on the other hand, has no known negative environmental effects. Secondly, many of the organics that have been used as PFAs are limited in size and shape. This is not a limitation for a system which uses ice particles as a PFA.

Ice particles have potential cost advantages over other PFA materials in a manufacturing environment. The elimination of an additional burn-out schedule saves time and money. Ice particles can easily be made in house as needed, eliminating the need to store PFAs in a warehouse before being used.

While porous ceramics are the focus of this study, this process also has the potential to be used to create porous metallic components via a metallic powder processing method. Because this is a new technique, many options must be investigated to find the ideal processing method. In this study, different ice shapes and sizes, ice volumes, setting and drying techniques, and sintering schedules are explored. The goal is to introduce this method as a feasible processing route to porous ceramics and to identify the boundaries of the process. It is expected that pore size, shape, interconnectivity, and percent porosity will be easily controlled using this novel fabrication technique.

MATERIALS AND METHODS

In order to illustrate the flexibility of this process, pores were made using ice particles of different shapes and sizes. Three kinds of ice particles, differing in size, shape, and uniformity were made for this study- crushed ice particles, small spherical particles, and large spherical particles. Crushed ice particles, random in size and shape, were made using a commercial blender. Small and large spherical ice particles were made using a water atomizer and dewar of liquid nitrogen.

Alumina slips were prepared using commercially available α-Al_2O_3 powder (99.9% purity, <1.0 micron, Alfa Aesar) and deionized water. Suspensions were prepared with an alumina solid loading of approximately 53 wt% to create a sufficiently thick ceramic slip. Ice particles were introduced to the alumina systems to create alumina/ice mixtures. Variations were made in the type of ice that was added (crushed, small spherical, or large spherical) as well as the amount of ice added by volume percent (0, 43, 56, or 75). Alumina/ice mixtures were immediately placed in a freezer after mixing at -5°C overnight to freeze.

Frozen samples were removed from the freezer and were left overnight on the lab bench to dry in room conditions. Once all ice had melted out and the samples were relatively dry to the touch, they were moved to a furnace for further drying. Complete drying was achieved in a furnace for 12 hours at 150°C. Samples were fired for 2 hours at either 1300°C, 1500°C, or 1600°C using a heating and cooling rate of 1.5°C/ min.

Fired samples were characterized to understand the effect of ice size, ice amount, and sintering temperature on porosity and microstructure. Archimedes' principle was utilized to qualitatively understand the amount of total, open, and closed porosity within samples. SEM and optical microscopy were used to quantitatively assess pore size and qualitatively review pore shape and degree of pore uniformity. Optical microscopy was additionally used to characterize the ice particles.

RESULTS AND DISCUSSION

Ice Particles

Figure 1 depicts optical images of the large and small spherical ice particles made for this experiment. Large ice particles averaged approximately 3 mm in diameter and all particles exhibited relatively decent uniformity in size and spherical shape as can be seen in the figure below. Small ice particles ranged in size from 10 to 200 microns in diameter and were less uniform in size.

Figure 1. Optical images of a) large and b) small spherical ice particles.

Porous Alumina Imaging

Figure 2 shows optical images of final fired alumina parts. Qualitative assessments were made on pore shape, size, uniformity, and distribution with these images. Clear differences are observed in the sizes of pores resulting from the large and small spherical ice particles. Figure 2a depicts a sample made from large ice particles. It's pores range anywhere from 0.5 to 2 mm. The pores visible in Figure 2b are significantly smaller, ranging from a few tens of microns up to 200 microns. The optical micrograph seen in Figure 2c is that of a control sample which was prepared without using any ice. While the surface of the sample is somewhat rough, no pores are visible.

Pores observed in the optical microscope images matched well with the ice particles used to create them. Pore widths observed in the optical images were slightly smaller than the sizes of the spherical ice particles, however the shape of the ice particles appears to have translated well into the ceramic sample. Sintering can account for the slightly smaller pore size. Pore collapse during drying also accounts for existing pores smaller than the PFA used to create them.

Figure 2. Optical images of alumina prepared with a) large spherical ice
b) small spherical ice and c) no ice.

SEM images were obtained and analyzed to more closely examine the presence of smaller pores in the samples. Figure 3 shows the porosity within a sample made with 43 vol% small spherical ice and sintered at 1500°C. Two different magnifications are used. Pores in these samples range in size from approximately 20μm to 200μm and are randomly distributed throughout the sample.

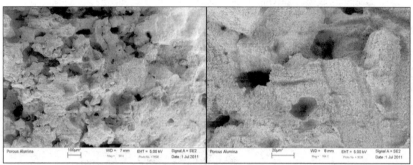

Figure 3. Large and small pores present in sample with 43 vol% small spherical ice, fired at 1500°C.

The sizes of pores visible in Figure 3 are consistent with the size range of the small spherical particles imaged in Figure 1. Pores seen in the SEM images tend to be relatively round in shape. Some elongated pores can be found in the sample. It is possible that these pores were formed by two spherical ice particles that were in contact with one another in the frozen green body. Pore interconnectivity was seen throughout the samples and was not well controlled in this experiment.

Archimedes Porosity Measurements

Figure 4 illustrates the changes in porosities (total, open and closed) versus the volume percent of ice that was added to the alumina slurry. Each of the samples presented in this figure underwent the exact same set of processing steps except for the amount of ice introduced to the slurry.

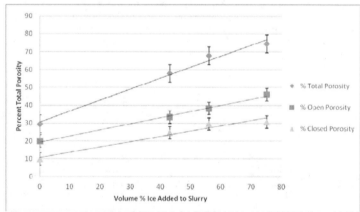

Figure 4. Percent porosity as a function of volume percent ice added to the slurry.

The presence of porosity in control samples which did not contain any ice indicates that there are other sources of porosity in this process besides the PFA added to the slurries. This porosity most likely arises from the packing density of the ceramic powder particles and sintering schedule used. In this study, the powders were not selected or processed such that they would form a fully dense body without the addition of a PFA. Porosity levels almost doubled between samples with 0 vol% ice and 43 vol% ice. These samples were processed in exactly the same way, and so this increase in porosity can only be accounted for by the addition of ice to the alumina slurry.

Figures 5 and 6 depict percent porosities (open, closed, and total) resulting from samples sintered at all three sintering temperatures. Samples in Figure 5 were prepared with blended ice and samples in Figure 6 were prepared with small spherical ice.

Particle transport and necking cause pores to close up during the solid-state sintering process. Figures 5 and 6 indicate a strong decrease in porosity as sintering temperatures increase. The size of the ice particles used as PFA's did not change the total amount of porosity observed in samples, however it did change the type of porosity. Samples that were prepared with smaller spherical shaped ice particles displayed higher amounts of closed porosity than samples which used the larger crushed ice particles. The smaller spherical ice particles were more likely isolated from one another within the green bodies and did not have as much interconnectivity to open the pores to the environment. It appears that larger ice particles are more likely to touch one another in a green body, producing higher amounts of open porosity.

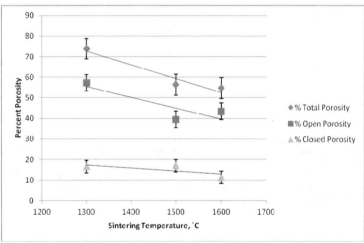

Figure 5. Porosity as a function of sintering temperature. Samples prepared with 43 vol% crushed ice particles.

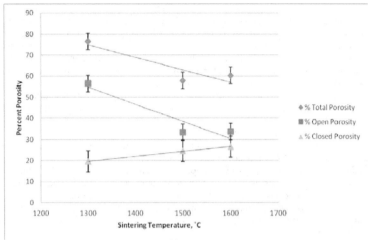

Figure 6. Porosity as a function of sintering temperature. Samples prepared with 43 vol% small spherical ice particles.

CONCLUSION

This study presented a new route to porous ceramics that has the potential to be low cost, environmentally friendly and allow for a high degree of control over the types of pores it creates. Porous alumina was successfully prepared with the novel use of ice particles as a pore forming agent. Alumina samples were prepared with pores ranging from 20μm to 3mm in diameter. Porosity levels ranged from 30% to 75%. Amount of porosity was controlled through the amount of ice particles added to the ceramic slurry as well as the firing temperature of the green body. The size and shape of ice particles were able to be changed based on the size and shape of the ice particles that were made.

While the total amount of porosity and the shape of the pores was able to be controlled within this study, more work will need to be done to address other factors that were not able to be controlled in this study. Interconnected porosity, for example, was randomly seen in various samples. This study aimed to prove the feasibility of using ice particles as pore forming agents. Although more work is needed to gain better control over porosity, the simplicity of this process makes it extremely appealing over other current fabrication methods.

REFERENCES

[1]L.N.W. Damoah and L. Zhang, AlF$_3$ reactive Al$_2$O$_3$ foam filter for the removal of dissolved impurities from molten aluminum: Preliminary results, *Acta Materialia*, **59**, 896-913 (2011).
[2]Y.M. Jo, R.B. Hutchison, and J.A. Raper, Characterization of ceramic composite membrane filters for hot gas cleaning, *Powder Technology*, **91**, 55-62 (1997).
[3]A. Julbe, D. Farrusseng, and C. Guizard, Porous ceramic membranes for catalytic reactors -- overview and new ideas, *J. of Membrane Science*, **181**, 3-20 (2001).
[4]D.J. Green, Structure and Properties of Porous Ceramics including Fibrous Insulation, *Encyc. Mat. Sci. and Tech.*, 7762-64 (2001).
[5]C.W. Tang, et al., Production of synthetic lightweight aggregate using reservoir sediments for concrete and masonry, *Cement and Concrete Composites*, **33**, 292-300 (2011).

[6]K. Rezwan, et al., Biodegradable and bioactive porous polymer/inorganic composite scaffolds for bone tissue engineering, *Biomaterials*, **27**, 3413-31 (2006).

[7]S.L. Chen, et al., Large pore heavy oil processing catalysts prepared using colloidal particles as templates, *Catalysis Today*, **125**, 143-8 (2007).

[8]Z.H. Wen, et al., Preparation of porous ceramics with controllable pore sizes in an easy and low-cost way, *Mat. Characterization*, **59**, 1335-38 (2008).

[9]N.J. Manjooran and G.R. Pickrell, Biologically self-assembled porous polymers, *J. Mat. Pro. Tech.*, **168**, 225-29 (2005).

[10]E. Gregorová and W. Pabst, Porous ceramics prepared using poppy seed as a pore-forming agent, *Ceram. Inter.*, **33**, 1385-88 (2007).

[11]E. Gregorová, et al., Porous alumina ceramics prepared with wheat flour, *J. Euro. Ceram. Soc.*, **30**, 2871-80 (2010).

[12]J. Bai, Fabrication and properties of porous mullite ceramics from calcined carbonaceous kaolin and [alpha]-Al_2O_3, *Ceram. Inter.*, **36**, 673-8 (2010).

[13]M. Dressler, et al., Burnout behavior of ceramic coated open cell polyurethane (PU) sponges, *J. Euro. Ceram. Soc.*, **29**, 3333-39 (2009).

RANDOM-HOLE OPTICAL FIBER SENSORS AND THEIR SENSING APPLICATIONS

Ke Wang, Brian Scott, Neal Pfeiffenberger and Gary Pickrell

Center for Photonics Technology and Department of Materials Science & Engineering
Virginia Tech, Blacksburg VA

ABSTRACT
    Silica optical fibers with a randomly varying porous cladding have been fabricated. These fibers, termed "Random-hole optical fibers (RHOFs)", have been utilized for optical sensors for various applications. Different sensor architectures have been fabricated in the RHOF. For example, long period gratings were fabricated in the RHOF by using a CO_2 laser through a point-by-point technique. In-fiber Fabry-Perot interferometers were fabricated by splicing a short section of multi-mode RHOF between two single-mode RHOFs. The long period grating sensors show an excellent sensitivity to external changes in refractive index, and thus can be utilized in chemical and biological sensing. The Fabry-Perot sensors, because of their short length, are ideal for most of the physical measurands.

INTRODUCTION TO RANDOM-HOLE OPTICAL FIBERS

    Microstructured Optical Fibers (MOFs), also termed as "Holey fibers", have attracted intensive theoretical and experimental investigations for telecommunication and sensing applications, since their first experimental demonstration in 1996 [1]. The first appearance of the MOF is based-on the ordered hole structures in the fiber cladding, which is often referred to as "Photonic Crystal Fibers" (PCFs). The PCFs are optical fibers whose claddings consist of an array of microstructured air holes running along the fiber axis which are arranged in a symmetric pattern, and therefore were referred to as photonic "crystal" fibers. Based-on the guiding mechanism, the PCFs can be categorized into: the high-index-guiding PCFs and the photonic bandgap PCFs [2]. The high-index-guiding PCFs have a solid core that has a higher refractive index than that of the cladding, where the cladding contains air holes causing a lower average effective refractive index. This type of PCF guides the light through the "Modified Total Internal Reflection". In contrast, the photonic bandgap PCFs have a lower refractive index core region such as a hollow (air) core. Such a structure creates a two-dimensional photonic crystal structure in the cladding and a defect in the core. The existence of the defect in the core region breaks the ordered structures of the photonic crystal and thus light is guided by the photonic bandgap mechanism, where light located in the stopband of the photonic crystal cannot propagate except in the core (defect) region.

    The PCFs are most commonly fabricated with a Stack-and-Draw technique [3]. The PCF preform is prepared by manually stacking an array of silica capillaries in a strict geometry to form the desired air-silica structure, around a solid central rod or a hollow tube. Then the preform is drawn using a fiber draw tower at temperature up to 2000°C. The uniformly sized holes (tubes) are symmetrically arranged around the central core region. A strict geometry of the structure is critical for the light guidance. However, it is almost unavoidable to break the ordered geometry in the fabrication process, which will largely impair the guiding properties.

    Recently, this problem has been overcome by developing of a novel fabrication technique [4, 5]. Instead of producing the ordered holes in the cladding by carefully stacking the silica capillaries in a strict geometry, the new technique produces a large number of holes in a variety of sizes, randomly distributed throughout the cladding, without the need for tube stacking. The holes in these fibers are made in-situ as the fiber preform is being drawn into fibers. The process is relatively simple and very cost effective. In addition, these new fibers do not suffer the same penalties for variations in symmetric ordering of the holes as a function of the length of the fiber. Since the new fibers have thousands of randomly distributed holes, variations in size or locations of a few of the holes does not change the overall properties observed. This type of fiber is commonly referred to as the Random Hole Optical Fiber (RHOF). The porosity of RHOFs is generated during the actual fiber drawing process. In one variation of this process, a preform is prepared in which a solid core is suspended in the middle of a fused silica tube, both made of pure silica. The space between the two layers is filled with silica powders mixed with a small amount of gas-producing materials, such as silicon nitride and/or silicon carbide. These gas-producing materials will release a gas (N_2 or CO_2), when oxidized at high temperatures. These materials when oxidized can leave only pure silica glass behind, as shown in Equations (1) and (2).

$$SiC + 2O_2(g) = SiO_2 + CO_2 \qquad (1)$$
$$SiN_4 + 3O_2(g) = 3SiO_2 + 2N_2 \qquad (2)$$

The heated gas, which is distributed throughout the preform, produces small bubbles, which results in randomly sized and spaced tubes in the final fiber laterally distributed in the cladding region. There are a sufficient number of holes to provide a consistent lowering of the refractive index. SEM micrographs of the RHOF are shown in Figure 1.

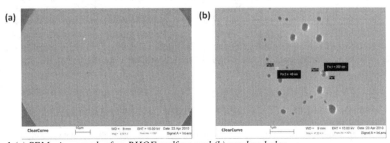

Figure 1 (a) SEM micrograph of an RHOF endface and (b) random holes.

RANDOM-HOLE OPTICAL FIBER SENSORS

Long period gratings in random-hole optical fibers

Long period gratings (LPGs), whose periodicities ranging from 100 to 1,000 μm, have been extensively studied and utilized as sensing devices [6]. The fabrication of LPGs in telecommunication fibers has utilized various techniques such as the UV laser, the CO_2 laser [7], the electric arc [8] and the mechanical method [9]. Writing LPGs with a CO_2 laser is simple and cost effective compared with the UV-writing technique such as the high cost of the UV laser, optical components, and the complexity of the mechanical setup. In addition, the CO_2 laser can be controlled more precisely in terms of pulse energy and duration, thus this technique can fabricate more consistent LPGs [10]. The LPG can be written to the fiber by a point-by-point manner until the fundamental core mode is coupled into strong cladding modes by the LPG. The fiber experiences periodic deformations produced by the CO_2 laser. The resonant wavelength λ satisfies equation (3) [6]:

$$\lambda = \left(n_{core} - n^m_{cladding}\right)\Lambda \qquad (3)$$

where n_{core} is the effective refractive index of the core; $n^m_{cladding}$ is the effective refractive index of mth-order resonant wavelength in the cladding; and Λ is the periodicity of the LPG.

The LPGs in the RHOF can be used to sense changes in the refractive index in the surrounding medium and therefore have significant potential to be developed into gas, chemical and biological sensors because of the sensitivity to the external refractive index [11].

In this research, LPGs are written in the RHOF by a CO_2 laser using the point-by-point technique. The fabrication setup was similar to what is previously reported for LPGs in regular telecommunication fibers [7, 10]. A 40-cm-long single-mode RHOF (Corning ClearCurve singlemode fiber) was fusion spliced onto regular Corning SMF-28 telecommunication fibers. Then the fiber was connected with a broad-band light source and the optical transmission was monitored by an optical spectrum analyzer. The fiber was fixed on a three-dimensional stage (Thorlabs PT3/M, resolution 2 µm) through two fiber clamps (Thorlabs T711-250). A 3-g weight was hung on the fiber in order to maintain a constant stress in the fiber while the grating was being written. The small weight is important in order to keep a constant strain in the fiber. A CO_2 laser (Synrad 48-2, wavelength 10.6 µm) was focused on to the RHOF through a cylindrical lens, whose focal length was 10 cm. The laser illumination time was precisely controlled by a computer program. The fabrication process was monitored by a CCD camera through an angle of 45 °.

During the fabrication process, the laser power and illumination time were manually selected by virtually observing the thermal effect on the fiber, where only slight deformation induced by the heat from the laser pulses was observed through the CCD camera. The laser should induce sufficient change in refractive index but should not seriously weaken the fiber's mechanical strength. Under the focusing by a 10-cm cylindrical lens, it was found that a suitable range was at 7% of the total power (25 W) and an illumination time of 3 s, such that LPGs can be successfully fabricated while maintaining sufficient mechanical strength. Under the laser illumination at such conditions, the fiber was slightly deformed. After the deformation, the fiber was moved to the next point (one periodicity) by manually tuning the stage. Thus, the LPG was fabricated by a point-by-point technique, until strong cladding mode couplings were observed in the optical spectrum analyzer.

In the experiment, the cladding mode coupling was normally observed after a few periods and started to reach the maximum coupling efficiency very quickly. For example, a typical grating transmission spectrum with a periodicity of 450 µm, is plotted. As it can be seen from Figure 2, the LPG reaches its maximum coupling efficiency of −9.81 dB at the wavelength 1498.5 nm with 14 periods. After reaching the maximum coupling efficiency, the transmission dips were saturated, with no further changes observed even with more periods of laser inscriptions.

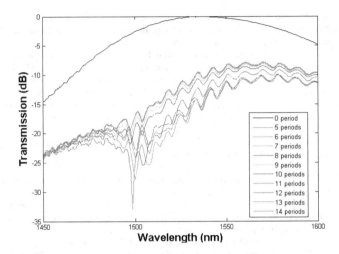

Figure 2. The transmission spectrum of an LPG in the RHOF with $\Lambda = 450$ μm.

In order to investigate sensing applications for refractive index, the LPG was immersed into standard refractive index liquids with refractive indices 1.400, 1.412, 1.420, 1.432 and 1.440, respectively. As shown in Figure 3, the LPG resonant wavelength shifts to lower wavelengths in response to the increase of the refractive indices in the surrounding mediums. This phenomenon is expected from Equation (3) because $n_{cladding}^m$ is increased by the larger refractive indices of the surrounding medium, and thus leads to shorter resonant wavelengths.

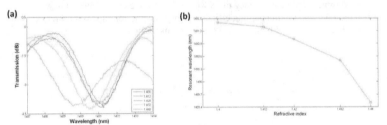

Figure 3. Resonant wavelength. (a) spectrums and (b) dips shift in response to the increase of refractive indices.

Intrinsic Fabry-Perot interferometric sensors

An intrinsic Fabry-Perot interferometric (IFPI) sensor is made by fusion splicing a section of multimode fiber between two single-mode fibers [12]. Because of the slight difference in the refractive indices of the two fibers, a Fabry-Perot cavity is formed and the reflectivity at each boundary can be expressed as:

$$R = \left(\frac{n_{mmf} - n_{smf}}{n_{mmf} + n_{smf}}\right)^2 \qquad (4)$$

The reflected spectrum from the IFPI sensor can be expressed as:

$$I(\lambda) = I_1(\lambda) + I_2(\lambda) + \sqrt{I_1(\lambda)I_2(\lambda)} \cos(\frac{2\pi \cdot OPD}{\lambda} + \varphi_0) \qquad (5)$$

Where I_1 and I_2 are the reflection from two boundaries, φ_0 is the initial phase and OPD=2nd; OPD stands for optical pathlength difference, n and d are the refractive index and the physical length of the multimode fiber, respectively.

An IFPI sensor in the RHOF can also be developed by fusion splicing a section of multimode RHOF (Corning ClearCurve multimode fiber) between two single-mode RHOFs. The fabrication process is similar to the process in reference [12]. A MicronOptics optical sensor interrogation system (Si725) was used to obtain the interference spectrum, as shown in Figure 4.

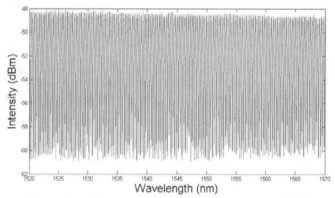

Figure 4. Interference spectrum from an IFPI sensor in the RHOF.

A temperature sensor can be developed by the IFPI sensor in the RHOF. The sensor was tested in a furnace from room temperature up to 250 °C. The temperature response with a linear fitting curve is shown in Figure 5 [13].

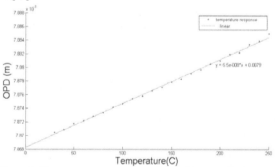

Figure 5. The temperature response of the IFPI sensor in the RHOF.

CONCLUSIONS

Random-hole optical fibers have been fabricated using a simplified and cost-effective technique by producing a large number of holes in the fiber cladding, randomly distributed in sizes from nano to micro meters. These fibers are also developed into sensing devices by writing long period gratings in them and also by making them into intrinsic Fabry-Perot interferometers. Refractive index and temperature sensing experiments were conducted to verify the sensing capabilities of these devices and they have shown promising performances.

REFERENCES

[1]   J. C. Knight, T. A. Birks, P. S. Russell, and D. M. Atkin, "All-silica single-mode optical fiber with photonic crystal cladding," *Optics Letters,* vol. 21, pp. 1547-1549, Oct 1 1996.

[2]   P. Russell, "Photonic crystal fibers," *Science,* vol. 299, pp. 358-362, Jan 17 2003.

[3]   P. S. Russell, T. A. Birks, J. C. Knight, and B. J. Mangan, "Photonic Crystal Fibre and A Method for Its Production," *US Patent 6985661,* 2006.

[4]   D. Kominsky, G. Pickrell, and R. Stolen, "Generation of random-hole optical fiber," *Optics Letters,* vol. 28, pp. 1409-1411, Aug 15 2003.

[5]   G. Pickrell, D. Kominsky, R. Stolen, F. Ellis, J. Kim, A. Safaai-Jazi, and A. B. Wang, "Microstructural analysis of random hole optical fibers," *Ieee Photonics Technology Letters,* vol. 16, pp. 491-493, Feb 2004.

[6]   R. Kashyap, *Fiber Bragg gratings,* 2nd ed. Burlington, MA: Academic Press, 2010.

[7]   D. D. Davis, T. K. Gaylord, E. N. Glytsis, S. G. Kosinski, S. C. Mettler, and A. M. Vengsarkar, "Long-period fibre grating fabrication with focused CO2 laser pulses," *Electronics Letters,* vol. 34, pp. 302-303, Feb 5 1998.

[8]   I. K. Hwang, S. H. Yun, and B. Y. Kim, "Long-period fiber gratings based on periodic microbends," *Optics Letters,* vol. 24, pp. 1263-1265, Sep 15 1999.

[9]   G. Rego, "Long-period fiber gratings mechanically induced by winding a string around a fiber/grooved tube set," *Microwave and Optical Technology Letters,* vol. 50, pp. 2064-2068, Aug 2008.

[10]   K. Wang and G. Pickrell, "Long Period Gratings in Random Hole Optical Fibers for Refractive Index Sensing," *Sensors,* vol. 11, pp. 1558-1564, Feb 2011.

[11]   G. Pickrell, W. Peng, and A. Wang, "Random-hole optical fiber evanescent-wave gas sensing," *Optics Letters,* vol. 29, pp. 1476-1478, Jul 1 2004.

[12]   Z. Y. Huang, Y. Z. Zhu, X. P. Chen, and A. B. Wang, "Intrinsic Fabry-Perot fiber sensor for temperature and strain measurements," *Ieee Photonics Technology Letters,* vol. 17, pp. 2403-2405, Nov 2005.

[13]   B. Qi, G. R. Pickrell, J. C. Xu, P. Zhang, Y. H. Duan, W. Peng, Z. Y. Huang, W. Huo, H. Xiao, R. G. May, and A. Wang, "Novel data processing techniques for dispersive white light interferometer," *Optical Engineering,* vol. 42, pp. 3165-3171, Nov 2003.

WETTING PROPERTIES OF SILICON INCORPORATED DLC FILMS ON ALUMINUM SUBSTRATE

Tae Gyu Kim[1]

Van Cao Nguyen[2]

Hye Sung Kim[3]

Soon-Jik Hong[4]

Ri-ichi Murakami[5]

[1]Dept. of Nanomechatronics Engineering, Pusan National University, Korea

[2]Department of Nano Fusion Technology, Pusan National University, Korea

[3]Dept. of Nano Materials Engineering, Pusan National University, Korea

[4]Division of Advanced Materials Engineering, Kongju National University, Korea

[5]Department of Mechanical Engineering, The University of Tokushima, Tokushima, Japan

ABSTRACT

In this study, we studied wetting properties of Si-DLC that are the important property in some applications in industries and researches. The Radio Frequency plasma-enhanced chemical vapor deposition (RF-PECVD) was used for making the Si-DLC film on the Aluminum substrates. Firstly, Si-DLC films are hydrophobic material. Contact angle of Si-DLC film was about 60°. However, it was changed when etched in O_2 plasma and heated at high temperature. The contact angle was decreased about 2° by not only oxygen plasma treatment for 10min but also by heat treatment at 70°C. In addition, an increase of heat treatment temperature makes the contact angle of Si-DLC film decrease. The contact angle, Micro Raman spectroscope and nano indentation was use for studying characteristics of Si-DLC films.

INTRODUCTION

Diamond-like carbon (DLC) is the name commonly accepted for hard carbon coatings, which have similar mechanical, optical, electrical and chemical properties to natural diamond, without dominant crystalline lattice structure instead. They are amorphous and consist of mixture of sp^3 which is diamond matrix and sp^2 bond graphite clusters embedded in an amorphous sp^3-bonded carbon matrix. Generally, it contains various quantities of hydrogen. Recently, DLC films have come to the center stage of developing coatings for moisture resistant lubricant, water repellant and cathode for lithium batteries. Hydrophilic nature of DLC films played an important role in the application fields mentioned [1-6].

EXPERIMENTALS

The principal aim of this work was to produce the change of wetting properties of Si-DLC thin films when treated in O_2 plasma and heated at high temperature. All these samples have been prepared using RF-PECVD system with a radio frequency standard of 13.56MHz, as shown in Figure 1. Coating conditions of DLC film on Aluminum substrates are shown in Table I. Aluminum substrates were cleaned in an ultrasonic bath containing acetone and ethanol for 10 min, respectively, and rinsed in distilled water. Before being placed in the coating chamber, substrates were dried in heater at 45°C. The substrates were further cleaned by Ar bombardment prior to deposition. DLC films were deposited using two different methods: a) Si-DLC and b) HMDS [hexamethyldisilane:$(CH_3)_3SiSi(CH_3)_3$)]. The Si-DLC film is normally deposited by PECVD system using CH_4 and SiH_4 gas mixture. The silane gas is known as an effective silicon source. But it is also a very dangerous material due to its high reactivity. In order to overcome this problem of silane, several novel silicon containing organic compounds were introduced for silicon reaction source to make Si-DLC films. In this study, HMDS was used as a precursor. HMDS is cheaper than the other organic silicon materials and easy to handle due to its chemical stability at room temperature. Since liquid HMDS has a considerably low vapor pressure at room temperature, the HMDS as a silicon source was introduced into the plasma chamber by bubbling HMDS contained cylinder with heated at a temperature of 40°C, which enables a sufficiently high vapor pressure to be obtained. DLC films were deposited on Aluminum substrates using thermal electron excited plasma CVD apparatus and $(CH_4 + HMDS)$ gas. Hydrophilic characteristic of Silicon incorporated DLC surfaces and HMDS surface has been studied by using O_2 plasma etching and heat treatment. The temperature of heat treatment was room temperature, 500°C, 600°C, 700°C for 10 min. Also, heat treatment of HMDS sample is done at 700°C for 20 min for comparison of characteristic with non-heated sample.

Table I. Conditions of Si-DLC and HMDS films on Aluminum substrates by RF-PECVD.

Specimen	Conditions			
	RF Power (W)	Deposition Time (minutes)	O_2 Plasma Treatment Time (minutes)	Heat Treatment Temperature (°C)
Si-DLC	600	30	5	500
Si-DLC	600	30	5	600
Si-DLC	600	30	5	700
HMDS	600	30	5	700

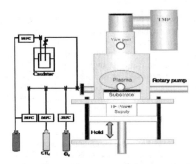

Figure 1. A schematic diagram of RF-PECVD deposition system.

The contact angle was measured immediately after O_2 plasma etching and heat treatment by a sessile drop method. In addition, Raman spectroscopy was used to analyze atomic bonds in the film.

RESULTS AND DISCUSSIONS

The wetting contact angles on the samples are shown in Table II and Figure 2, 3, and 4. The hydrophilic characteristic of silicon incorporated DLC surfaces has been studied by the use of O_2 plasma etching and heat treatment. The properties of DLC films were evaluated by various techniques including contact angle, Micro Raman spectroscope, scratch test and Nano indentation. The typical Contact angle of Si-DLC and HMDS films were shown Fig. 2. Contact angle of Si-DLC and HMDS films were 58.7° and 73.2°, respectively. Figure 3 shows the contact angle of Si-DLC and HMDS films after O_2 plasma etching. Contact angle of Si-DLC and HMDS films after O_2 plasma etching were 2.2° and 4.6°, respectively. However, 72 hours later, the contact angle of Si-DLC film was rapidly increased from 2.2° to 60° as shown in Figure 5 (a). To increase holding time, heat treatment had been done. Figure 4 shows the contact angle of Si-DLC and HMDS films after heat treatment ((a) Si-DLC 500°C, (b) Si-DLC 600°C, (c) Si-DLC 700°C, (d) HMDS 700°C). After 500°C, 600°C and 700°C heat treatment of Si-DLC, contact angle were 30.4°, 8.9° and 2.4°, respectively. And contact angle was 10.3° after heat treatment of HMDS film at 700°C. In addition, contact angle of Si-DLC and HMDS films heat treated at 700°C were maintained below 25° for 480 hours and 362 hours, respectively, as shown in Figure 5 (b) and (c). Also, the increase of heat treatment temperature makes the contact angle of Si-DLC film decreases [2]. It is considered that hydrophilic surface can be maintained because DLC is crystallized by heat treatment.

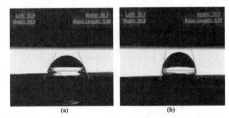

(a)   (b)

Figure 2. Contact angle of the (a) as-deposited Si-DLC [58.7°], (b) HMDS [73.2°].

(a)   (b)

Figure 3. Contact angle of the (a) Si-DLC [2.2°], (b) HMDS [4.6°] after O_2 plasma etching.

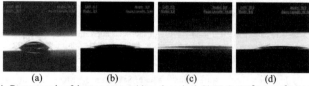

Figure 4. Contact angle of heat treatment (a) as-deposited Si-DLC, 500°C [30.4°], (b) 600°C [8.9°], (c) 700 °C [2.4°] and (d) HMDS, 700 °C [10.3°].

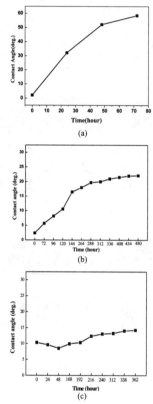

(a)

(b)

(c)

Figure 5. The change of contact angle of Si-DLC films after O₂ plasma etching treated for 72 hours (a), and after heat treated for 480 hours (b); HMDS films after heat treatment at 700°C for 362 hours (c).

Table II. Results of wetting contact angle of Si-DLC and HMDS films.

Sample	Contact angle (°)
As-deposited Si-DLC	58.7
O2 plasma etching treated Si-DLC	2.2
Heat treated Si-DLC (500°C)	30.4
Heat treated Si-DLC (600°C)	8.9
Heat treated Si-DLC (700°C)	2.4
O$_2$ plasma treated HMDS	73.2
Heat treated Si-DLC (700°C)	10.3

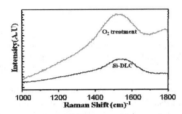

Figure 6. Raman spectra of Si-DLC film after treatment and before treatment.

The typical Raman spectrum of the amorphous DLC film was shown in Fig.6, the peaks at about 1550cm^{-1} were assigned to amorphous DLC. DLC is exist after O$_2$ treatment [4].

Figure 7. Optical observation of the scratch on Si-DLC film.

Figure 7 shows the result of scratch test of Si-DLC sample. A Revetest Scratch Tester (CSM Instruments, Switzerland) equipped with a 200 μm Rockwell-C tip was employed. In the scratch test a stylus is dragged across the film surface at a constant speed, progressively increasing the normal load. At each point, the lateral force is measured. Preliminary tests in the systems were carried out in order to identify the load range at which fracture occurs. Based on those tests, the following conditions were selected for the scratch test. Initial load (normal) of $F_{N,0} = 0N$; final load of $F_{N,f} = 50N$; Loading rate (normal load) $F = 50N/min$. Sliding speed $v = 10mm/min$. These conditions result in a total scratch length of $L = 10mm$. Optical micrographs of scratch damage to the sample scratched at maximum load of 50mN, critical normal and lateral forces at cracking of 35 N.

The hardness of the coating was determined by nano-indentation. Nanoindentation of the systems was carried out in a Nano-indentation (CSM Instruments, Switzerland) equipped with a Berkovich tip. In order to measure the elastic modulus of the films, the Oliver and Phar method was used . Applied loads were lower than 4.5mN so that depths did not exceed 10% of film thicknesses. Hardness of the DLC film was measured to be 2306 Hv.

SUMMARY
    In this study, the chemical composition of DLC films and the interface region of the DLC/Al substrate have been analyzed with the measurement of contact angle, Raman spectra analysis, scratch test and nano indentation. Si-DLC coating were deposited by using RF-PECVD system. Contact angle of Si-DLC film was about 60°. After oxygen plasma treatment for 10min, the contact angle was about 2°. When the heat treatment is done at 700°C, the sample has very low contact angle of about 2°. Contact angle of DLC films heat treated at 700°C were maintained below 25° for 480hours (Si-DLC films) and 15° for 362 hours (HMDS films). Also, the increase of heat treatment temperature makes the contact angle of Si-DLC film decrease. It is confirmed that the peak at about 1550 cm^{-1} is an amorphous DLC peak from the Raman spectroscopy analysis. Si-DLC film deposited using RF-PECVD technique shows high hardness of 2306Hv. The excellent film adhesion even at higher loads is obtained from Si-DLC coating.

REFERENCES
[1]R. Paul, S. Dalui, S. N. Das, R. Bhar, A. K. Pal, Hydrophobicity in DLC films prepared by electrodeposition technique, *Applied Surface Science*, **255**, 1705–1711 (2008).
[2]Y. Yin, L. Hang, J. Xu, D. R. McKenzie, M. M. M. Bilek, Surface adsorption and wetting properties of amorphous diamond-like carbon thin films for biomedical applications, *Thin Solid Films*, **516**, 5157–5161 (2008).
[3]Midathada Anil, Sk. Faruque Ahmed, Jin Woo Yi, Myoung-Woon Moon, Kwang-Ryeol Lee, Yu Chan Kim, Hyun Kwang Seok, Seung Hee Han, Tribological performance of hydrophilic diamond-like carbon coatings on Ti–.6Al–.4V in biological environment, *Diamond and Related Materials*, **19**, 300-304 (2010).
[4]H. Schulz, M. Leonhardt, H. J. Scheibe, B. Schultrich, Ultra hydrophobic wetting behavior of amorphous carbon films, *Surface and Coatings Technology*, **200**, 1123-1126 (2005).
[5]J.C. Damasceno, S.S. Camargo Jr., DLC-SiO$_x$ nanocomposite films deposited from CH$_4$: SiH$_4$: O$_2$ gas mixtures, *Surface and Coatings Technology*, **200**, 6279-6282 (2006).
[6]M. Schvartzman and S. J. Wind, Plasma fluorination of diamond-like carbon surfaces: mechanism and application to nanoimprint lithography, *Nanotechnology*, **20**, 145306-145313 (2009).

NANOPOROUS AG PREPARED BY ELECTROCHEMICAL DEALLOYING OF MELT-SPUN CU-AG-SI ALLOYS

Guijing Li, FeiFei Lu, Linping Zhang, Zhanbo Sun*, Xiaoping Song, Bingjun Ding, Zhimao Yang

(MOE Key Laboratory for Non-equilibrium Synthesis and Modulation of Condensed Matter, State Key Laboratory for Mechanical Behavior of Materials, Xi'an Jiaotong University, Xi'an 710049, P. R.

* Corresponding Author.   szb@mail.xjtu.edu.cn

ABSTRACT

Nanoporous Ag ribbons were prepared by electrochemical dealloying of the melt-spun $Cu_{80-X}Ag_{20}Si_X$ (X=5, 10, 15, in mol%) alloys in 0.5M $CuSO_4$ aqueous solution at room temperature. The Cu and Si elements in the precursor alloys could be removed almost during dealloying. The pores and ligaments of the dealloyed Cu-Ag-Si alloys distributed throughout the cross section of the ribbons. The nanoporous Ag were refined evidently with the addition of Si in the precursor alloys. The pores of the nanoporous Ag prepared from the $Cu_{70}Ag_{20}Si_{10}$ alloy were refined to ~ 40 nm and the ligaments with an average size of ~100 nm showed granular shape. The nanoporous Ag exhibited strong enhancement of the surface-enhanced Raman scattering (SERS) due to the refinement of microstructures.

INTRODUCTION

Nanoporous bulk metals with high surface area and ultralow density have recently attracted increasing attention [1-2] in the last decade. The nanoporous Ag prepared by dealloying were extensively studied due to its unique optical properties [3], electrochemical performance [4], catalytic activity [5-6] and so on. The essence of dealloying was a selective corrosion. In the process, one or more elements in the precursory alloys were removed, and the nanoporous structure would be obtained [7]. The intrinsic properties of a nanoporous metal are mainly dependent on the morphologies, porosity, pore and ligament sizes [8-9]. Therefore, many studies focused on the selection of precursory alloys to investigate the nanoporous Ag. Among the investigations, many alloys, such as Mg-Ag [10], Ag-Al [11], Ag-Zn [12] and Cu-Ag [13] were used as the precursory alloys to investigate nanoporous Ag. The results suggested that the compositions of the precursory alloys had a significant influence on the nanoporous Ag.

In addition, great efforts have been devoted to the dealloying process. Xu et al [14] investigated the preparation and structure evolution of nanoporous Ag by carefully tuning the experimental parameters during dealloying of Ag-Al alloys in aqueous NaOH or HCl solutions. Luechinger et al [15] tuned the microstructure of the nanoporous Ag by exposure to low pH and elevated temperatures. Our previous work prepared nanoporous Ag ribbons by electrochemical corrosion of the melt-spun Cu-Ag alloys [16]. It was found that the structures of the nanoporous Ag were strong dependent on the microstructures of the precursory alloys, and the nanoporous Ag exhibited different enhancements of the surface-enhanced Raman scattering (SERS) effect. These results proposed that the control of the microstructures of the precursory alloy was very important to obtain desirable nanoporous Ag.

For an alloy with given composition, the microstructures were often refined by an addition of the special component and increase of solidification rate. In addition, the dealloying mechanism of precursor alloys with multicomponents has not been studied well. The present investigation reported an approach to prepare and control the nanoporous Ag through electrochemical dealloying of the melt-spun Cu-Ag-Si ribbons. The effect of the Si contents in precursory alloys on the morphology and the surface-enhanced Raman scattering of the nanoporous Ag were discussed. The dealloying mechanisms of the Cu-Ag-Si alloys with different phase constitutions were investigated.

EXPERIMENTAL

The Cu-Ag-Si alloys were prepared from pure Ag (99.99%), pure Cu (99.99%) and pure Si (99.99%) in a vacuum arc furnace at an Ar atmosphere. The ingots with a mass of 6g were obtained. After the ingots inserted into a quartz tube, they were heated by high frequency induction to an appropriate temperature. The melt-spun Cu-Ag-Si ribbons were achieved through a single roller melt spinning at the speed of 30 m/s^{-1} under a pressure of 5×10^3 Pa Ar gas. The width and thickness of the ribbons were approximately 1.5~2 mm and 30~40 μm. The phase structures of the as-quenched ribbons were analyzed use a Bruker D8 advanced X-ray diffractometer. The microstructures of the as-quenched ribbons were observed by a JSM-7000F microscope (SEM) equipped with a backscattered electron detector. The as-quenched ribbons were used to prepare nanoporous Ag by electrochemical dealloying in a beaker of 0.5M $CuSO_4$ aqueous solution. The dealloying voltage was provided by a HH1732C5 potentiostat (the setting value was 0.05V). The Cu-Ag-Si ribbons served as the anode and a silver rod as the cathode.

The morphology and size characteristics of the as-dealloyed ribbons were analyzed by the microscope. The surface-enhanced Raman scattering effects of the obtained nanoporous Ag ribbons were measured by Laboratory Ram JY-HR800 spectrometer after they immersed in the 0.1mM Rhodamine 6G solutions for 20min (dried in air). The output laser power was 17mw/100 (D2 filter) and the operating wavelength was 632.8nm.

RESULTS AND DISCUSSION

Figure 1 shows the X-ray diffraction (XRD) patterns of the melt-spun Cu-Ag-Si alloys. The $Cu_{75}Ag_{20}Si_5$ ribbons consisted of two phases: fcc-Cu and fcc-Ag. As the Si content increased to 10%, the alloy was composed of fcc-Cu, fcc-Ag and Cu-Si compounds. On further increasing the Si content ($Cu_{65}Ag_{20}Si_{15}$), the relative intensity of the Cu-Si compounds peaks became stronger while the intensity of Cu peaks weakened. These results confirmed that Si contents had significant influence on the phase constitutions of the melt-spun Cu-Ag-Si alloys.

The microstructures of the melt-spun Cu-Ag-Si ribbons are shown in Figure 2. The brighter was Ag-rich and the matrix was Cu-rich. For the binary $Cu_{80}Ag_{20}$ alloy, as shown in Figure 2(a), a typical divorced eutectic microstructure was present. The coarse Ag-rich phase distributed along the Cu-rich grain boundaries were with a rod-like shape. As 5% Si added into the alloy, the Ag-rich phase was refined although the rod-like shape could still be observed, see Figure 2(b). For the melt-spun $Cu_{70}Ag_{20}Si_{10}$ ribbons, the Ag-rich phase was granular, and the size was very fine and uniform, as shown in Figure 2(c). As the Si content further increased to 15%, the size of the Ag-rich

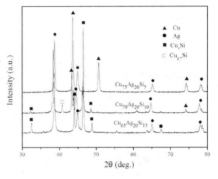

Figure 1. XRD patterns of the melt-spun $Cu_{75}Ag_{20}Si_5$, $Cu_{70}Ag_{20}Si_{10}$ and $Cu_{65}Ag_{20}Si_{15}$ ribbons.

particles was not uniform and the larger Ag particles, as shown in Figure 2(d), were coarser than that in the $Cu_{70}Ag_{20}Si_{10}$ ribbons. The results indicated that the size of Ag particles were dependent on Si content in the melt-spun Cu-Ag-Si ribbons, where, they were finest in the $Cu_{70}Ag_{20}Si_{10}$ ribbons.

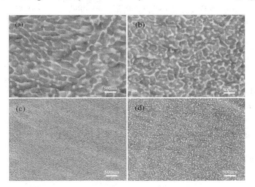

Figure 2. Backscattered electron images of the melt-spun $Cu_{80}Ag_{20}$ (a), $Cu_{78}Ag_{20}Si_5$ (b), $Cu_{70}Ag_{20}Si_{10}$ (c) and $Cu_{65}Ag_{20}Si_{15}$ ribbons

According to the Ag-Cu and Ag-Si binary phase diagram [17-18], the solubility of Cu in fcc-Ag and Ag in fcc-Cu are about 5% and 13% at the eutectic temperature, respectively. However, the solubility of Si in fcc-Ag is very low and no intermetallic compound forms. The effect of rapid solidification on solubility enhancement would be obvious infinitesimal in the Si-Ag and Cu-Ag systems. The microstructures indicated that the Ag-rich phase in $Cu_{75}Ag_{20}Si_5$ ribbons (Fig. 2b) formed during the solidification and the fcc-Ag was supersaturated solid solution of Cu and Si, resulting in the smaller lattice parameters. From the microstructure (Fig. 2c), it was suggested that the fine Ag particles were precipitation as the Si content was 10%. There were few Cu atoms dissolved in fcc-Ag and the lattice parameters were closed to pure Ag. It was deduced that the Ag particles in the $Cu_{65}Ag_{20}Si_{15}$ ribbons also solidified during melt spinning due to further increase of Si content, the fcc-Ag was supersaturated and resulted in the Ag particles coarsening and enhancement of solubility.

Table 1 lists the EDS results of the Cu-Ag-Si ribbons dealloyed at the potential of 0.05 V in 0.5M $CuSO_4$ solution for different time. It is very clearly that the Si and Cu could not be detected in the dealloyed alloys by EDS. These results indicated that the pure Ag ribbons have been achieved by electrochemical dealloying the melt-spun Cu-Ag-Si alloys. Especially, it was noted that the increase of the Si content in the alloys promoted the dealloying due to the shorter dealloying time.

Table 1. EDS results of the nanoporous Ag prepared by electrochemical dealloying of the $Cu_{80-x}Ag_{20}Si_X$ (X=5, 10, 15) alloy at the potential of 0.05 V for different time.

Samples	Cu (at.) %	Ag (at.) %	Si (at.) %	Dealloying Time (h)
$Cu_{75}Ag_{20}Si_5$	0	100	0	8
$Cu_{70}Ag_{20}Si_{10}$	0	100	0	5
$Cu_{65}Ag_{20}Si_{15}$	0	100	0	2

Figure 3 shows the plan-view of the dealloyed $Cu_{80}Ag_{20}$ and $Cu_{80-x}Ag_{20}Si_X$ (X=5, 10, 15) ribbons. It can be observed from the Figure 3(a) that the pore and ligament sizes in the dealloyed $Cu_{80}Ag_{20}$ ribbons were larger than 250 nm. The ligament morphology was mainly rod-like. For the nanoporous Ag prepared from the $Cu_{75}Ag_{20}Si_5$ alloy, the pore channels, as shown in Figure 3(b), have an average size of ~ 150 nm thought the morphology of the nanoporous Ag was not changed much. The ligament morphology in the dealloyed $Cu_{70}Ag_{20}Si_{10}$ exhibited a particle-like shape and the pore channels were less than 40 nm, as show in Figure 3(c). The distribution of the pores and ligaments was very uniform. These results indicated that the pore size was considerably decreased as the increase of the Si contents less than 10% in the precursory alloy. As Si content increased to 15%, the pore and ligament sizes of nanoporous Ag, as shown in Figure 3(d), decreased further. However, many finer nano Ag particles with diameter of ~10 nm distributed among the ligaments, which resulted in a reduction of porosity. These results clearly indicated that the nanoporous Ag were obtained by electrochemical dealloying of the melt-spun Cu-Ag-Si alloys and the morphology and size of the nanoporous Ag could be controlled by the Si contents in the Cu-Ag alloys.

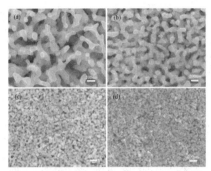

Figure 3. Plan-view SEM images of nanoporous Ag prepared by electrochemical dealloying of the melt-spun $Cu_{80}Ag_{20}$ (a), $Cu_{75}Ag_{20}Si_5$ (b), $Cu_{70}Ag_{20}Si_{10}$ (c) and$Cu_{65}Ag_{20}Si_{15}$ (d) ribbons at the potential of 0.05V in 0.5 M $CuSO_4$.

Figure 4 illustrates the cross-sectional view of the nanoporous Ag ribbons prepared from the Cu-Ag-Si alloys. From Figure 4(a), (c) and (e), it can be observed that the pores distributed across the entire section of the dealloyed ribbons. The morphology of the interconnected Ag ligaments prepared from $Cu_{75}Ag_{20}Si_5$ alloy, as shown in Figure 4(b) was short rod-like shape. As

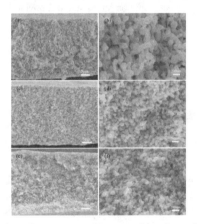

Figure 4. Section-view of nanoporous Ag prepared by electrochemical dealloying of the melt-spun $Cu_{75}Ag_{20}Si_5$ (a)-(b), $Cu_{70}Ag_{20}Si_{10}$ (c)-(d) and $Cu_{65}Ag_{20}Si_{15}$ (e)-(f) alloys at the potential of 0.05v in the 0.5 M $CuSO_4$ solution.

the Cu-Ag alloys contained 10% Si, the ligaments in nanoporous Ag were particle shape. The average sizes of the ligaments and pore channels were less than 90 nm and 40 nm, respectively, see Figure 4(d).

As the Si content increased to 15%, the pore size reduced obviously. But the granular ligaments, as shown in Figure 4(f), were obviously non-uniform. It was noticed that many particles distributed among the larger ligaments, which agree with the plan morphology.

The Cu-Ag-Si system would consist of fcc-Cu, fcc-Ag and Cu-Si compounds depending on the compositions. The $Cu_{75}Ag_{20}Si_5$ alloy was composed of Ag-rich and Cu-rich solid solutions (Figure 1), but the size of Ag-rich phase was decreased due to the addition of Si (Figure 2). During dealloying, the Ag-rich phase evolved into the ligaments of nanoporous Ag after the Cu-rich phase was decomposed. As a result, the nanoporous Ag prepared from the melt-spun $Cu_{75}Ag_{20}Si_5$ alloy had a smaller size than that from the $Cu_{80}Ag_{20}$ (Figure 3(a)-(b)) due to the refinement of microstructure. As the Si content increased to 10%, the Ag-rich phase further refined, as a result, the sizes of the pores and ligaments became much smaller after the Cu-rich phase and Cu-Si compound were decomposed. As for the $Cu_{65}Ag_{20}Si_{15}$ alloy, excess Si caused that the size of the Ag particles in the precursor alloy was very inhomogeneous (Figure 2(d)). After the Cu-Si compounds incorporated Ag atoms decomposed, the crystallized smaller Ag particles distributed among the larger ligaments. As a result, the ligament size in the nanoporous Ag became non-uniform and had lower porosity (Figure 4(f)). These results indicated that the Si content in the Cu-Ag alloys should be an appropriate amount so as to obtain the refined nanoporous Ag. In addition, when the Si and Cu atoms dissolved in the Ag-rich particles were extracted, the Ag atoms could crystallize attaching the primary Ag, which had very little effect on the morphology of the nanoporous structures according to our prior work [16].

Figure 5 shows the Raman spectra of the nanoporous Ag prepared from the melt-spun Cu-Ag-Si alloys immersed in 10^{-4} M R_6G aqueous solution for 20 min. It is obviously that the

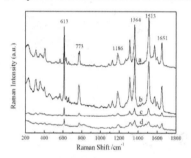

Figure 5. SERS spectra of the dealloyed Cu-Ag-Si alloys. a, b, c and d indicate the SERS effect of nanoporous Ag prepared from the melt-spun Cu75Ag20Si5, Cu70Ag20Si10 , Cu65Ag20Si15 and Cu80Ag20 alloys.

nanoporous Ag had stronger surface-enhanced Raman intension signals of R_6G, and the SERS effect was enhanced with the increase of Si from 5at.% to 10at.% in the precursory alloys. However, the SERS effect of the nanoporous Ag prepared from the $Cu_{65}Ag_{20}Si_{15}$ alloy were weakened. Studies showed that the intensity of SERS signals for nanoparticles were stronger than that for nanorods [19]. The morphology of nanoporous Ag prepared from the $Cu_{75}Ag_{20}Si_5$ alloy was rod-like shape, while that of the dealloyed $Cu_{70}Ag_{20}Si_{10}$ alloy was particle-like. Therefore, nanoporous Ag prepared from the

$Cu_{70}Ag_{20}Si_{10}$ alloy had remarkable SERS effect. Thought the pores size of the dealloyed $Cu_{65}Ag_{20}Si_{15}$ was reduced, the porosity of the nanoporous Ag ribbons was clearly lower than that from $Cu_{75}Ag_{20}Si_5$ and $Cu_{70}Ag_{20}Si_{10}$ (Figure 4). More short-distance hot-spots which contributed to the enhancement of the surface-enhanced Raman scattering could not be obtained [20], which resulted in the weakness of the SERS signals.

CONCLUSION

In summary, the nanoporous Ag with different morphology and pore sizes were successfully prepared by electrochemical dealloying of the melt-spun Cu-Ag-Si alloys at the potential of 0.05V in 0.5 M $CuSO_4$ solution. The pore channels of nanoporous Ag were reduced from 250 nm to 150 nm due to adding 5% Si into the precursor alloys. With the increasing of Si content from 5% to 10%, the sizes of pores/ligaments of nanoporous Ag was further reduced to about 40 nm. However, as the Si content was increased to 15% ($Cu_{65}Ag_{20}Si_{15}$), the size of the Ag particles in the precursor alloy was very inhomogeneous, which caused non-uniform of nanoporous Ag. In addition, the Si in the precursory alloys promoted the dealloying. Nanoporous Ag exhibited stronger enhancement of the SERS effect with the addition of 5% and 10% Si in the precursor alloys. However, the SERS effect was weaken due to the decrease of porosity for the dealloyed $Cu_{65}Ag_{20}Si_{15}$ alloy. The present work could present a method to control the nanoporous metals by adding of sacrifice elements into the precursory alloys.

ACKNOWLEDGEMENTS

This work is sponsored by Natural Science Foundation of China (50871081, 51071116, 50901056, 50871080) and the National 863 Program Projects of China (2009AA03Z320)

REFERENCES

[1] B. C. Tappan, S. A. Steiner III and E. P. Luther, Nanoporous metal foams, Angew. Chem. Int. Ed., 49, 4544-4565 (2010).

[2] S. Polarz and B. J. Smarsly, Nanoporous materials, Nanosci. Nanotechno., 6, 581- 612 (2002).

[3] Y.Y. Li and Y. Ding, Porous AgCl/Ag nanocomposites with enhanced visible light photocatalytic properties, J. Phys. Chem. C, 114, 3175-3179 (2010).

[4] F. Jia, C. F. Yu, K. J. Deng and L. Z. Zhang, Nanoporous metal (Cu, Ag, Au) films with high surface area: general fabrication and preliminary electrochemical performance, J. Phys. Chem. C, 111, 8424-8431 (2007).

[5] H. Ji, J. Frenzel, Z. Qi, X.G. Wang, C. C. Zhao, Z.H. Zhang and G. Eggeler, An ultrafine nanoporous bimetallic Ag-Pd alloy with superior catalytic activity, CrystEngComm.,12, 4059-4062 (2010).

[6] Z. Liu, L. Huang, L. Zhang, H. Ma and Y. Ding, Electrocatalytic oxidation of d-glucose at nanoporous Au and Au-Ag alloy electrodes in alkaline aqueous solutions, Electrochimica Acta, 54, 7286-7293 (2009).

[7] J. Erlebacher, M. J. Aziz, A. Karma, N. Dimitrov and K. Sieradzki, Evolution of Nanoporosity in Dealloying, Nature, 410, 451 (2001).

[8] A. Wittstock, V. Zielasek, J. Biener, C. M. Friend and M. Bäumer, Nanoporous Gold Catalysts for Selective Gas-Phase Oxidative Coupling of Methanol at Low Temperature, Science, 327, 319-322 (2010).

[9] Arne Wittstock, Jürgen Biener and Marcus Bäumer, Nanoporous gold: a new material for catalytic and sensor applications, Phys. Chem. Chem. Phys., 12, 12919-12930 (2010).

[10] H. Ji, X. G. Wang, C. C. Zhao, C. Zhang, J. L. Xu and Z. G. Zhang, On the vacancy-controlled dealloying of rapidly solidified Mg-Ag alloys, CrystEngComm., 13, 2617 (2011).

[11] Z. Zhang, Y. Wang, Z. Qi, W. Zhang, J. Qin, and J. Frenzel, generalized fabrication of nanoporous metals (Au, Pd, Pt, Ag, and Cu) through chemical dealloying, J. Phys. Chem. C, 113, 12629-12636 (2009).

[12] F. H. Yeh, C. C. Tai, J. F. Huang, I. W. Sun, Formation of porous silver by electrochemical alloying/dealloying in a water-Insensitive zinc chloride-1-ethyl-3-methyl imidazolium chloride ionic liquid, J. Phys. Chem. B, 110, 5215 (2006).

[13] R. Morrish and A. J. Muscat, Nanoporous silver with controllable optical properties formed by chemical dealloying in supercritical CO_2, Chem. Mater., 21, 3865-3870 (2009).

[14] C. Xu, Y. Li, F. Tian and Y. Ding, Dealloying to nanoporous silver and its implementation as a template material for construction of nanotubular mesoporous bimetallic nanostructures, ChemPhysChem., 11, 3320-3328 (2010).

[15] N. A. Luechinger, S. G. Walt and W. J. Stark, Printable nanoporous silver membranes, Chem. Mater., 22, 4980-4986 (2010).

[16] G. J. Li, X. P. Song, Z. B. Sun, S. C. Yang, B. J. Ding, S. Y., Z. M. Yang and F. Wang, Nanoporous Ag prepared from the melt-spun Cu-Ag alloys, Solid State Sci., 13, 1379-1384 (2011).

[17] W.J. Boettinger, D. Shechtman, R.J. Schaefer and F. S. Biancaniello, The Effect of Rapid Solidification Velocity on the Microstructure of Ag-Cu Alloy, Metallurgical transactions A 15, 55-66 (1984).

[18] R.W. Oleslnskl, A.B. Gokhale and G.J. Abbaschian, the Ag-Si system, Bulletin of Alloy Phase Diagrams 10, 635- 639 (1989).

[19] V. S. Tiwari, T. Oleg, G. K. Darbha, W. Hardy, J. P. Singh and P. C. Ray, Non-resonance SERS effects of silver colloids with different shapes, Chem. Phys. Lett., 446, 77-82 (2007).

[20] S. Y. Lee, L. Hung, G. S. Lang, J. E. Cornett, I. D. Mayergoyz, O. Rabin, Dispersion in the SERS Enhancement with silver nanocube dimers , ACS Nano., 4, 5763-5772 (2010).

EFFECT OF FILM THICKNESS ON ELECTRICAL AND OPTICAL PROPERTIES OF ZNO/AG DUAL LAYER FILM

Hiromi Yabe(1), Eri Akita(1), Pangpang Wang(2), Daisuke Yonekura(1), Ri-ichi Murakami(1), and Xiaoping Song(2)
(1)Department of Mechanical Engineering, the University of Tokushima, Tokushima, Japan
(2)Department of Materials Physics, School of Science, Xi'an Jiaotong University, Xi'an, China

ABSTRACT
    ZnO/Ag dual layer films were deposited on glass substrates by DC magnetron sputtering method to clarify the effect of film thickness on the electrical and optical properties. Both the Ag film and the ZnO film were deposited with different thickness. The electrical and optical properties of the dual layer film were mainly changed by the Ag film thickness. The sheet resistance of 2.3 ohm/sq with transmittance over 80% in the visible light was obtained under the deposition time of 30 seconds for Ag and 30 minutes for ZnO film. From the result of SEM observations, the Ag was deposited on the glass substrate as discontinuous structure. Therefore, ZnO/Ag dual layer film will be expected for good performance in the transparent electrode.

INTRODUCTION
    Transparent conductive oxide (TCO) material is defined as the transmittance over 80% and the resistivity less 10^{-3} ohm*cm. TCO is applied in many thin film devices, such as solar cells, touch panel screens, and flexible displays [1-4]. Now, indium tin oxide (ITO) film is widely used for transparent conductive oxides [5-6]. However, ITO is limited to use the application because it is unstable chemically and thermally in various environments. In addition, indium is expensive and is unevenly distributed in the Earth. Therefore, it is necessary to develop low-cost TCO materials with high properties instead of ITO.
    ZnO is the most attractive TCO material as an alternative to ITO [7]. TCO materials based on ZnO have good transmittance and electrical conductivity, which are low-cost, abundant amount, non-toxic [8-9], and having high stability in a hydrogen plasma [10-11]. However, their electrical conductivity is rather lower in some cases when they adopt as a TCO for a practical product. To improve the conductivity of ZnO film, multilayer structure has been examined, which consists of three or five layers of metal and semiconductor or dielectrics materials. Some authors reported that symmetric ZnO/Ag/ZnO multilayer films have high transparency and conductivity and can improve both the optical transmittance and electrical conductivity of ZnO films [12-17]. While multilayered films with three layers are reported in some cases, few reports have focused on depositing a thin intermediate Ag layer onto a glass substrate directly, and have studied the properties of asymmetric ZnO/metal films. Thick silver layer deposited on the glass substrate shows low light transmission and high reflection. However, the transmittance spectra keep high in visible range when the silver layer is thin. So the optimization of film thickness of Ag and ZnO layers is important to achieve high conductivity and transmittance simultaneously. In this work, ZnO/Ag dual layers are deposited on the glass substrate by DC magnetron sputtering method. The effect of each film thickness on electrical and optical properties of the dual layer film is examined.

PROCEDURE

ZnO/Ag dual layer films were deposited layer by layer on a glass substrate using a DC magnetron sputtering system (NACHI, SP-1530-1). The base pressure was approximately 3.0×10^{-3} Pa and the sputtering process was performed at a chamber pressure of approximately 7.0×10^{-1} [Pa] without heating. A pure silver target (>99.99%) was used for depositing the Ag layer under an argon atmosphere with a mass flow rate of 50 sccm. A pure zinc target (>99.99%) was used to deposit the ZnO layer in a mixed argon/oxygen atmosphere with 50 sccm of argon gas and 3 sccm of oxygen gas after Ag layer deposition. Table 1 shows the sputtering conditions that were used to prepare the dual layer films. The layer thickness was controlled by the sputtering time. The Ag sputtering time was set to 10, 20, 30, 40, 50, and 60 seconds, and the sputtering time of ZnO layer was 10, 20, 30, 40, 50, 60, 70, 80, and 90 min. The deposition rate of silver layer was 12 nm/min. The deposition rate of the ZnO layer was as low as 0.5 nm/min. However, the thickness of each layer was not uniformly [18-19], therefore, each sample was distinguished by sputtering time. In this work, in order to observe the morphology of Ag nano layer by scanning electron microscope (SEM, Hitachi S-4700), the Ag was sputtered on glass substrates without coating the top ZnO layer.

The optical transmittance spectrum of the dual layer films was measured by a UV-vis-NIR spectrometer (JASCO, V570) in the wavelength range from 300 to 800 nm. The electrical properties were measured by a Hall effect measurement system (EDK, HEM-2000) following the Van der Pauw method. The structure of the dual layer films was investigated by X-ray diffraction (XRD, Bruker, D8 Advance). Surface roughness was determined and surface images were observed by atomic force microscope (AFM, Seikoh SPA300).

TABLE 1 Sputtering condition for preparation of ZnO/Ag dual layer films.

Layer type	Ar flow (SCCM)	O$_2$ flow (SCCM)	Target Current (A)	Target Voltage (V)	Sputtering time	Deposition rate
Ag	50	0	0.4	280	10,20,30,40,50,60 sec	12 nm/min
ZnO	50	3	0.1	250	10,20,30,40,50,60,70,80,90 min	0.5 nm/min

RESULTS AND DISCUSSION

Figure 1 shows the typical XRD patterns of the ZnO/Ag dual layer films as a function of the Ag and ZnO sputtering time. The peak intensity of Ag (111) increased with increasing Ag sputtering time, and that of ZnO (002) was stable as shown in Fig. 1 (a). In addition, the intensity of ZnO (002) increased with increasing ZnO sputtering time, while the intensity of Ag (111) was stable as shown in Fig. 1 (b). The results mean that amount of deposition of each layer increases on the glass substrate as the sputtering time increases.

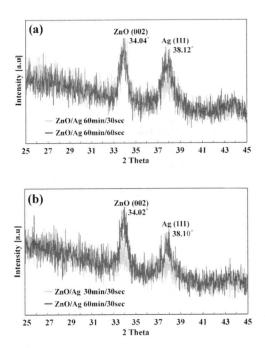

Figure 1 XRD patterns for ZnO/Ag dual layer films with different sputtering time.

Figure 2 shows the optical transmittance spectra of the ZnO/Ag dual layer films on the glass substrate as a function of Ag and ZnO sputtering time. The transmittance of the dual layer films decreased with increase of Ag sputtering time as shown in Fig. 2 (a). The maximum transmittance for Ag sputtering time 10 to 30 sec was higher than 80%. However, the transparency decreased, especially for longer wavelength, when the deposition time of Ag layer increases. At Ag sputtering time of 60 sec, the lowest transmittance of dual layer was obtained.

The transmittance slightly increased when the sputtering time of the ZnO layer increased as shown in Fig. 2 (b). The wavelength of the maximum transmittance for ZnO/Ag films were at λ= 380, 400, 470, and 550 nm for the respective ZnO layer sputtering time of 10, 30, 60, and 90 min. It means that with increasing ZnO sputtering time, the increased transmission represents a redshift from short to long wavelength in the visible region.

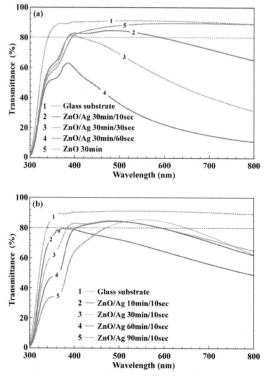

Figure 2 Optical transmittance spectra for the ZnO/Ag films on glass substrate.

To investigate the morphology of the Ag layer grown on the glass substrate, SEM and AFM observation was performed. Figure 3 shows the SEM images of the Ag layer deposited on the glass substrate for each Ag sputtering time without ZnO top layer. Figure 3 (a) shows a typical low magnification image of the sample surface. The Ag was deposited on the glass substrate locally. The locally-sputtered Ag films expand on the glass substrate with increasing Ag sputtering time, and then, Ag covers the substrate completely over 50 sec sputtering time. Figure 3 (b) – (g) shows high magnification images of each Ag sputtering time. The morphology of Ag layer changed with increasing Ag sputtering time. At the Ag sputtering time of 10 and 20 sec, Ag grains showed a discontinuous structure. On the basis of the transmittance of the ZnO/Ag dual layer films, the discontinuous Ag grains on the glass substrate will lead to keep the transmittance because the light can pass through the openings of the discontinuous structure. However the transmittance will decrease clearly for the Ag sputtering time over 30 sec, because the Ag layer becomes continuous structure and thick. From the SEM surface observations, it is found that the behavior of Ag layer from a discontinuous structure to a continuous structure is a key factor in optimizing the performance of dual layer films.

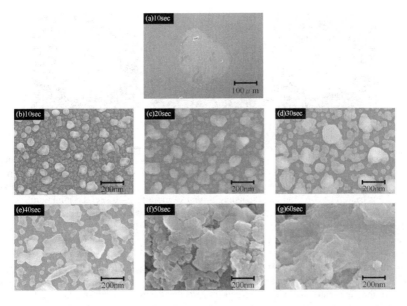

Figure 3 Top view SEM images of Ag films with different Ag sputtering time of 10 to 60 sec.

Figure 4 shows the AFM images of Ag single layer at the different Ag sputtering time on the glass substrate. The surface roughness of Ag single layer films increased with increasing the Ag sputtering time. Figure 5 shows the AFM images of dual layer at Ag sputtering time of 60sec and different ZnO sputtering time. The ZnO sputtering time had a little influence on the surface roughness of the ZnO/Ag dual layer film. The results mean that the reduction of transparency for the ZnO/Ag dual layer films is mainly caused by light scattering and reflecting on the surface with rough and thick layer

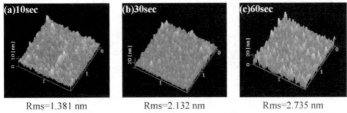

Rms=1.381 nm   Rms=2.132 nm   Rms=2.735 nm

Figure 4 The AFM images of the Ag single layer at different Ag sputtering time.

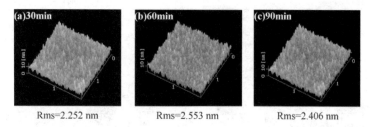

Rms=2.252 nm   Rms=2.553 nm   Rms=2.406 nm

Figure 5 The AFM images of the dual layers of Ag 60sec at different ZnO sputtering time.

Figure 6 shows the sheet resistance of ZnO/Ag dual layer films as a function of the Ag sputtering time. The resistivity curve shows two regions with different slopes. In the region corresponding to the sputtering time of less than 20 sec, a rapid drop of sheet resistance was observed. On the other hand, in the region corresponding to that of more than 20 sec, the minimum resistivity was about 9.72×10^{-6} ohm*cm at ZnO/Ag: 60min/60sec, which was close to the bulk Ag resistivity ($\sim$ 1.6×10^{-6} ohm*cm, at 300K) . The significant segmentation of the sheet resistance is due to the formation process of the Ag layer. At the beginning of deposition, the Ag grain is grown as islands on the glass substrate. When the sputtering time increases, the distance between each Ag grain is decreased, as shown in Fig.3. Therefore, the electrical conductivity of the dual layer films increases as shown in Fig. 6. In addition, there is quite small change for the resistivity of the dual layer films with increase of the ZnO sputtering time. The result suggests that the thickness of ZnO layer in the ZnO/Ag films has little influence on electrical properties of the dual layer films, and the structure of Ag layer is very important to improve the electrical properties and to keep the transparency.

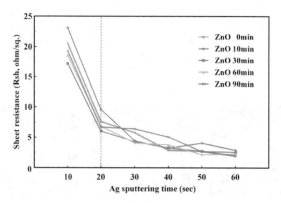

Figure 6 Sheet resistance of ZnO/Ag dual layer films with different sputtering time.

CONCLUSION

In conclusion, high transmittance ZnO/Ag dual layer films with low resistivity were obtained that are promising candidates for TCO applications. The effect of film thickness on electrical and optical properties of ZnO/Ag dual layer film was investigated.

1. The transmittance of ZnO/Ag dual layer films deposited on the glass substrate was enhanced close to 90% by decreasing Ag sputtering time and increasing ZnO sputtering time.

2. The morphology of the silver layer deposited on the glass substrate was changed from discontinuous structure to continuous structure when the sputtering time increased. The electrical conductivity of the dual layer films was influenced greatly by the morphological changes of the silver layer. An obvious change of resistivity was observed at the Ag sputtering time of 20 sec, and the ZnO/Ag films showed the lowest sheet resistance of 2.3 ohm/sq for Ag sputtering time of 60 sec.

3. The ZnO/Ag dual layer film with ZnO 30 min and Ag 30 sec can be optimized to have sheet resistance of 4 ohm/sq and a transmittance over 80% in the visible range. This makes it possible to fabricate the transparent conductive oxide with high performance at room temperature without using the heated substrate and annealing method.

REFERENCE

[1]A. J. Morfa, K. L. Rowlen, Appl. Phys. Lett, 92, 013504 (2008)

[2]S. Pillai, K. R. Catchpole, T. Trupke, M. A. Green, J. Appl. Phys. 101, 093105 (2007)

[3]N. Fang, H. Lee, C. Sun, X. Zhang, SCIENCE, vol.308 (2005)

[4]D. K. Gifford, D. G. Hall, Appl. Phys. Lett, 81, 23 (2002)

[5]C. Guillen, J. Herrero, Optics Communications, 282, 574 (2009)

[6]G. Leftheriotis, P. Yianoulis, D. Patrikios, Thin Solid Films, 306, 92 (1997)

[7]T. Mouet, T. Devers, A. Telia, Z. Messai, V. Harel, K. Konstantinov, I. Kante, M. T. Ta, Applied Surface Science, 256, 4114 (2010).

[8]I-Lun Hsiao, Yuh-Jeen Huang, Science of the Total Environment 409, 1219 (2011)

[9]Xiaohui Penga, Shelagh Palma, Nicholas S. Fisher, Stanislaus S. Wong, Aquatic Toxicology 102 186 (2011)

[10]N.H. Al-Hardan, M.J. Abdullah, A. Abdul Aziz, International Journal of hydrogen energy, 35, 4428 (2010)

[11]Zhen Zhou , K. Kato, T. Komaki,M. Yoshino, H. Yukawa,M. Morinaga, International Journal of hydrogen energy, 29, 323 (2004)

[12]P. Wang, D. Zhang, D. H. Kim, Z. Qiu, L. Gao, R. Murakami, X. Song, J. Appl. Phys. 106(1-5), 103104 (2009).

[13]D. Zhang, P. Wang, R. Murakami, X. Song, Appl. Phys. Lett. 96, 233114 (2010).

[14]D. R. Sahu, J. L. Huang, Thin Solid Films, 515, 876 (2006).

[15]D. R. Sahu, S. Y. Lin, J. L. Huang, Thin Solid Films, 516, 4728 (2008).

[16]D. R. Sahu, S. Y. Lin, J. L. Huang, Applied Surface Science, 252, 7509 (2006).

[17]A. I. Ievtushenko, V.A. Karpyna, V. I. Kazorenko, G. V. Lashkarev, V. D. Khranovskyy, V. A. Baturin, O. Y. Karpenko, M. M. Lunika, K. A. Avramenko, V. V. Strelchuk, O. M. Kutsay, Thin Solid Films, 2009

[18]S. H. Mohamed, Journal of Physics and Chemistry of Solids, 69, 2378 (2008).

[19]V. Sittinger, F. Ruske, W. Werner, C. Jacobs, B. Szyszka, D. J. Christie, Thin Solid Films, 516, 5847 (2008).

Author Index